T0319615

THE LIMITS OF FABRICATION

IDIOM INVENTING WRITING THEORY

Jacques Lezra and Paul North, series editors

THE LIMITS OF FABRICATION

MATERIALS SCIENCE, MATERIALIST POETICS

NATHAN BROWN

Fordham University Press *New York* *2017*

Fordham University Press also publishes its books in a variety of electronic formats. Some content that appears in print may not be available in electronic books.

Visit us online at www.fordhampress.com.

Library of Congress Cataloging-in-Publication Data available online at catalog.loc.gov.

Printed in the United States of America

19 18 17 5 4 3 2 1

First edition

A book in the American Literatures Initiative (ALI), a collaborative publishing project of NYU Press, Fordham University Press, Rutgers University Press, Temple University Press, and the University of Virginia Press. The Initiative is supported by The Andrew W. Mellon Foundation. For more information, please visit www.americanliteratures.org.

for Cynthia

Things, and present ones, are the absolute conditions.

—Charles Olson, "Equal, That Is, to the Real Itself" (1958)

Blt **o** by Bolt
Every single P
art is a crown
to Anatom

—Caroline Bergvall, *Goan Atom* (2001)

Work nano, think cosmologic.

—Shanxing Wang, *Mad Science in Imperial City* (2005)

CONTENTS

FIGURES

THE LIMITS OF FABRICATION

PROLOGUE: LIMITS

In the class-A clean room of the National Nanofabrication Laboratory in the heart of Silicon Valley, under the ultrahigh-magnification AFM (Atomic Force Microscope), I thumbed through every page of the whole collection of poetry books stolen from the Public Library, and I found only dots, dotted straight lines, dotted arcs. Abstract geometrical entities banging my dilated eyeballs.

—Shanxing Wang, *Mad Science in Imperial City* (2005)

A scene of reading: in the nanofabrication laboratory, we study the contents of the library. Our protocols of reading are materialist, empirical. Flipping through volumes of poetry, we examine their pages through the mediation of an Atomic Force Microscope: a device capable of nanoscale resolution, enabling the inspection and characterization of materials at molecular and submolecular scales of less than one-billionth of a meter. Not only the words and their referents but also the graphemes themselves, the ink of which they are composed, the fibers of the paper on which they are inscribed give way before our eyes onto "dots, dotted straight lines, dotted arcs." Concrete inscriptions give way onto "abstract geometrical entities," which themselves turn out to be obtrusively material, "banging my dilated eyeballs." At this scale, concrete things and abstract signs are indiscernibly *entities*, and it is the objectifying gaze of the laboratory technician that is accosted by the materials under inspection, not the other way around. Reading and writing, observation and fabrication, ideality and materiality encounter one another through the mediation of organs and instruments that magnify the fact of their complicity. In Shanxing Wang's 2005 volume of poetry, *Mad Science in Imperial City*, "nano" is a prominent signifier of that complicity, a prefix that appears 17 times across the 132 pages of the volume. In my epigraph from Wang's book, the nanofabrication laboratory is a site in which close reading is at once a scientific and a critical practice, in which poetic form becomes a matter of material structure, in which the materiality of the signifier is situated in the larger field of materiality per se. Nanotechnology is the organon of this situation. Taking our epigraph as a guide, let's begin to explore the situation it describes by following Wang's reader into the laboratory, the library.[1]

LIMIT 1: TECHNOSCIENCE

In a famous 1959 address to the American Physical Society, "There's Plenty of Room at the Bottom," Richard Feynman posed what he called "the final question" facing materials fabrication: "whether, ultimately—in the great future—we can arrange the atoms the way we want; the very atoms, all the way down!"[2] Feynman's "invitation to enter a new field of physics," in which he addressed "the problem of manipulating and controlling things on a small scale," is routinely cited as an inaugural impetus to the development of nanotechnology, the instrumental manipulation of matter at molecular and submolecular scale levels. Feynman's talk featured speculative proposals for positioning single atoms, building miniature computers, developing better electron microscopes, and engineering nanoscale machines (devices measuring around a billionth of a meter) that could perform mechanical tasks autonomously in "a billion tiny factories." But it was also a talk about reading and writing. "How do we write small?" Feynman asked. He proposed a method for inscribing a copy of the complete Encyclopedia Britannica on the head of a pin, and he concluded his talk by offering an award of $1,000 to the first person "who can take the information on the page of a book and put it on an area 1/25,000 smaller in linear scale in such manner that it can be read by an electron microscope."[3] That challenge was met in 1985 when Stanford graduate student Tom Newman programmed an electron beam apparatus to inscribe the first page of *A Tale of Two Cities* in the appropriate dimensions.[4] Each of the letters in Newman's text, however, was the width of approximately fifty atoms. A text composed at the scale of Feynman's limit-case scenario, through the direct positioning of single atoms, would require another approach to writing small, one that begins to elide the difference between writing and reading altogether.

In 1981, Gerd Binnig and Heinrich Rohrer at IBM laboratories in Zürich invented the Scanning Tunneling Microscope (STM), the first in a class of proximal probe microscopes to which the Atomic Force Microscope (AFM) mentioned in the epigraph also belongs. Binnig and Rohrer would share the 1986 Nobel Prize in Physics for their invention, which circumvented the resolution limitations of optical and electron microscopes through the use of a "tactile" interface, enabling the precise three-dimensional resolution of nanoscale matter, including single atoms. The STM, explains physicist Donald Eigler, "forms an image in a way which is similar to the way a blind person can form a mental image of an object by feeling the object."[5]

The basis of this tactile sensitivity is the quantum mechanical phenomenon of electron tunneling, in which an electron tunnels through a barrier or across a gap between proximate atoms. When a tiny metallic conducting needle, narrowing to a single atom at its tip, is brought into proximity with a conducting surface, electrons "tunnel" between the atomic tip and the atoms of the surface. This establishes a current, and since the magnitude of that current is minutely sensitive to the distance between the tip and the sample, it can be used to establish a feedback loop that adjusts the position of the tip in accordance with the atomic topography of the sample. Mounted on a piezoelectric transducer that adjusts its height with finite control, the tip is scanned across a surface, rising or falling according to the atomic terrain it encounters. As it moves, the position of the tip is measured and converted into a digitally mediated visual map of the sample's atomic structure, which is displayed on a monitor. The Atomic Force Microscope operates in a similar way, though the position of its tip can be calibrated according to a variety of physical forces existing at the interface—rather than quantum tunneling—including mechanical forces applied through direct contact with the sample (Fig. 1).

Crucially, both the STM and the AFM can also operate in a positioning mode, enabling users to manipulate single atoms deposited on a surface. When the tip of the microscope is lowered to the point at which the forces between the tip and the atom are sufficient to overcome those between the atom and the surface upon which it rests (but not so great as to "pick up" the atom entirely), the atom can be dragged or pushed across the surface with precision control. The resulting change in its position can then be recorded by returning the tip to its initial height and reimaging the sample (Figs. 2 and 3).

In 1989, Donald Eigler and Erhard Schweizer of IBM Almaden Labs answered Feynman's "final question" by demonstrating the STM's ability to manipulate single atoms in a way that foregrounded its capacities as a writing machine. Depositing thirty-five xenon atoms onto a substrate of nickel in an ultrahigh vacuum, cooled to liquid helium temperature, Eigler and Schweizer used an STM to arrange "the very atoms, all the way down" into an apocalyptic advertise-

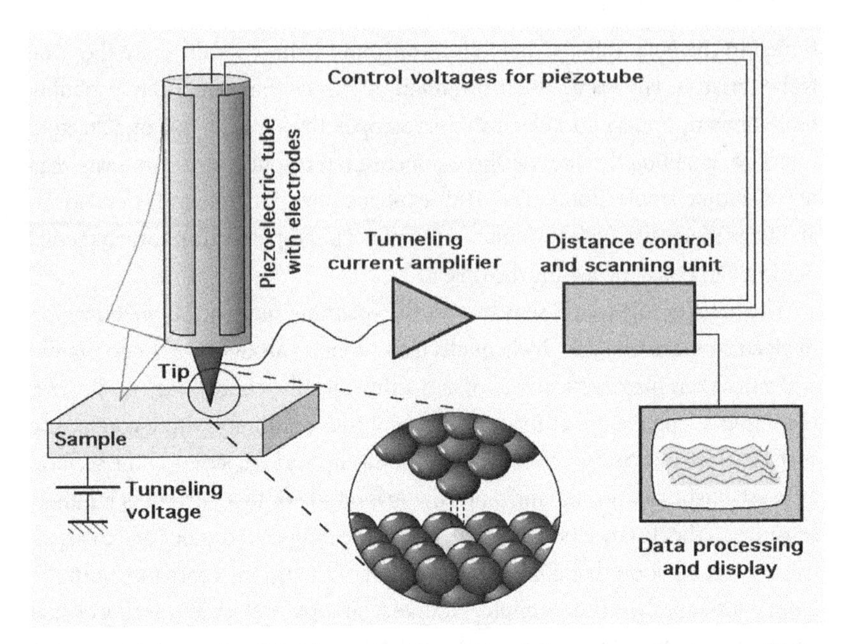

Figure 1 Schematic of the Scanning Tunneling Microscope. Courtesy of Michael Schmid, Institute of Applied Physics, Vienna University of Technology.

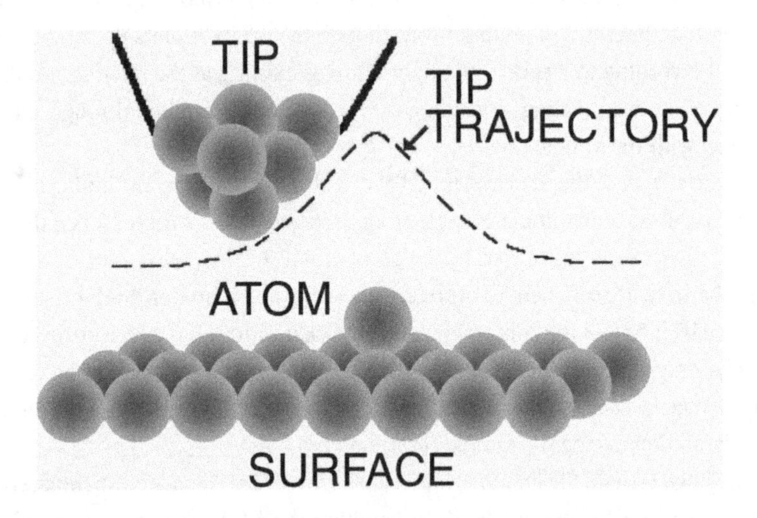

Figure 2 Scanning Tunneling Microscope in imaging mode. Courtesy of University of Wisconsin, Madison, Materials Research Science and Engineering Center.

ment for International Business Machines, inscribed onto the clean slate of what Feynman had called "the great future" (Fig. 4).

Noting that "it should be possible to assemble or modify certain molecules in this way," Eigler and Schweizer declared in their 1990 report to *Nature* that "the possibilities for perhaps the ultimate in device miniaturization are evident."[6] At the intersection of reading and writing, text and image, matter and meaning, scientific experimentation and corporate branding, they claimed that technoscience had arrived at the limits of fabrication.

LIMIT 2: POETRY

Around 1863, Emily Dickinson begins a poem by stating a problem: "I cannot live with You –." She then proceeds to unfold a labyrinth of grammatical, theological, and affective implications before arriving at an ambivalent solution:

> So We must meet apart –
> You there – I – here –
> With just the Door ajar
> That Oceans are – and Prayer –
> And that White Sustenance –
> Despair –[7]

"You there – I – here –": the first thing to notice about this line is that, along with Dickinson's characteristic dashes, it is composed entirely of deictic terms, or shifters. The dash is a minimal graphemic unit—pen touching down on paper with an instant's pressure, leaving the barest trace of furtive contact. Shifters are the piezoelectric transducers of grammar—minutely sensitive to the voltage of voice, expanding to generate an apparent fusion of body, language, world at the interface of the tongue's tip: "there."

The class of shifters includes pronouns ("You" and "I" in this case) as well as indices of spatio-temporal location: "there" and "here" in Dickinson's line but also "now" and "then" or "this" and "that." Roman Jakobsen explains that shifters are "distinguished from all other constituents of the linguistic code solely by their compulsory reference to the given message."[8] In other words, shifters have the indexical function of establishing an existential relationship between a particular subject and object, or place and time, through a particular speech act. As Emile Benveniste puts it, shifters constitute "an ensemble of 'empty' signs that are non-referential with respect to 'reality.' These signs are always available and become

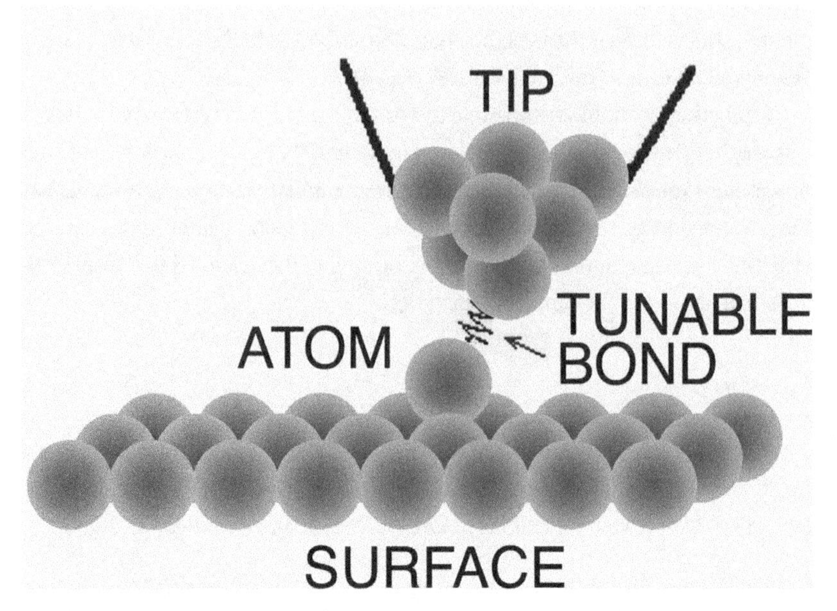

Figure 3 Scanning Tunneling Microscope in manipulation mode. Courtesy of University of Wisconsin, Madison, Materials Research Science and Engineering Center.

'full' as soon as the speaker introduces them into each instance of his discourse."[9] In his chapter on sense-certainty in the *Phenomenology of Spirit*, Hegel famously addresses the implications of these "empty" signs for written language as follows:

> To the question: "What is now?", let us answer, e.g. "Now is Night." In order to test the truth of this sense-certainty a simple experiment will suffice. We write down this truth; a truth cannot lose anything by being written down, any more than it can lose anything through our preserving it. If *now, this noon*, we look again at the written truth we shall have to say that it has become stale.[10]

"Here" and "now," the spoken shifter anchors a subject in a context. Later, reactivated by an act of reading, the context of the written shifter shifts. Some 170 years after Hegel's meditation, Roland Barthes would seize upon this slippage between spoken and written shifters as an instrument of the "destruction of the Author."[11]

If, in the course of reading Dickinson's poem, one relinquishes the impulse to ask after the identity of the "I" that is speaking, or the "You" with whom that "I" cannot live, that's because it becomes increasingly clear that the poem might as well be "about" the status of pronominal reference itself. Stanzas eleven and

Figure 4 Atomic IBM logo, 1989. Composed of thirty-five xenon atoms on a substrate of nickel. Image originally created by IBM corporation.

twelve, for example, seem at least as concerned with the interpersonal and semiotic dynamics of deixis as they are with the vagaries of the afterlife:

> And were You lost, I would be –
> Though My Name
> Rang loudest
> On the Heavenly fame
>
> And were You – saved –
> And I – condemned to be
> Where You were not –
> That self – were Hell to Me –

In the first line, "I" is exposed as a meaningless category without some "You" to whom it can be spoken, and the second line confirms the hollowness of proper names in the absence of the personal pronouns that link us with language. Benveniste points out that "consciousness of self is only possible if it is experienced by contrast. I use *I* only when I am speaking to someone who will be *you* in my address. . . . It is this condition of dialogue that is constitutive of *person*." "Language is possible," he continues,

> only because each speaker sets himself up as a subject referring to himself as *I* in his discourse. Because of this, *I* posits another person, the one who, being as he is com-

> pletely exterior to "me," becomes my echo to whom I say *you* and who says *you* to me. This polarity of persons is the fundamental condition of language.[12]

In other words, if "*I*" were "condemned to be / Where *You* were not – / That self – were Hell to *Me*."

The ambivalent solution to the problem posed by Dickinson's poem—to "meet apart"—thus preserves the exteriority of "You" while saving "Me" from a solipsism bereft of language. In his 1976 collection, *Shifters*, Steve McCaffery sums up this cut that connects not so much as a "polarity of persons" but as a situation of suspended sublation, absorbing differentiation and inclusion into a dialectic at a standstill:

> you
> are what
> i
>
> am apart
> from
>
> what
> i
>
> is
> a part
>
> of [13]

"You there – I – here –": without reference to Dickinson, McCaffery homes in on the undercurrents of her line in a note to his volume. "Shifters shift," he writes, "within a topography and topology of text where every 'i' is a 'here' and every 'you' a 'there'. poems then of openness and closure. semiotic bars and semiotic centres unfolding as tests of their own meanings."[14]

"Semiotic bars" returns us to Dickinson's dashes (the logic and function of which the whole poem might be taken to explicate), opening a passage to the second feature of her line that I want to highlight. "You there – I here –" discloses its most radical semiotic qualities not through its grammar but only when we grasp its paragrammatic operation. Leon Rodiez explains that a text is paragrammatic "in the sense that its organization of words (and their denotations), grammar, and syntax is challenged by the infinite possibilities provided by letters or phonemes combining to form networks of significance not accessible through conventional reading habits."[15] In Dickinson's line, the paragram operates on a scale below that

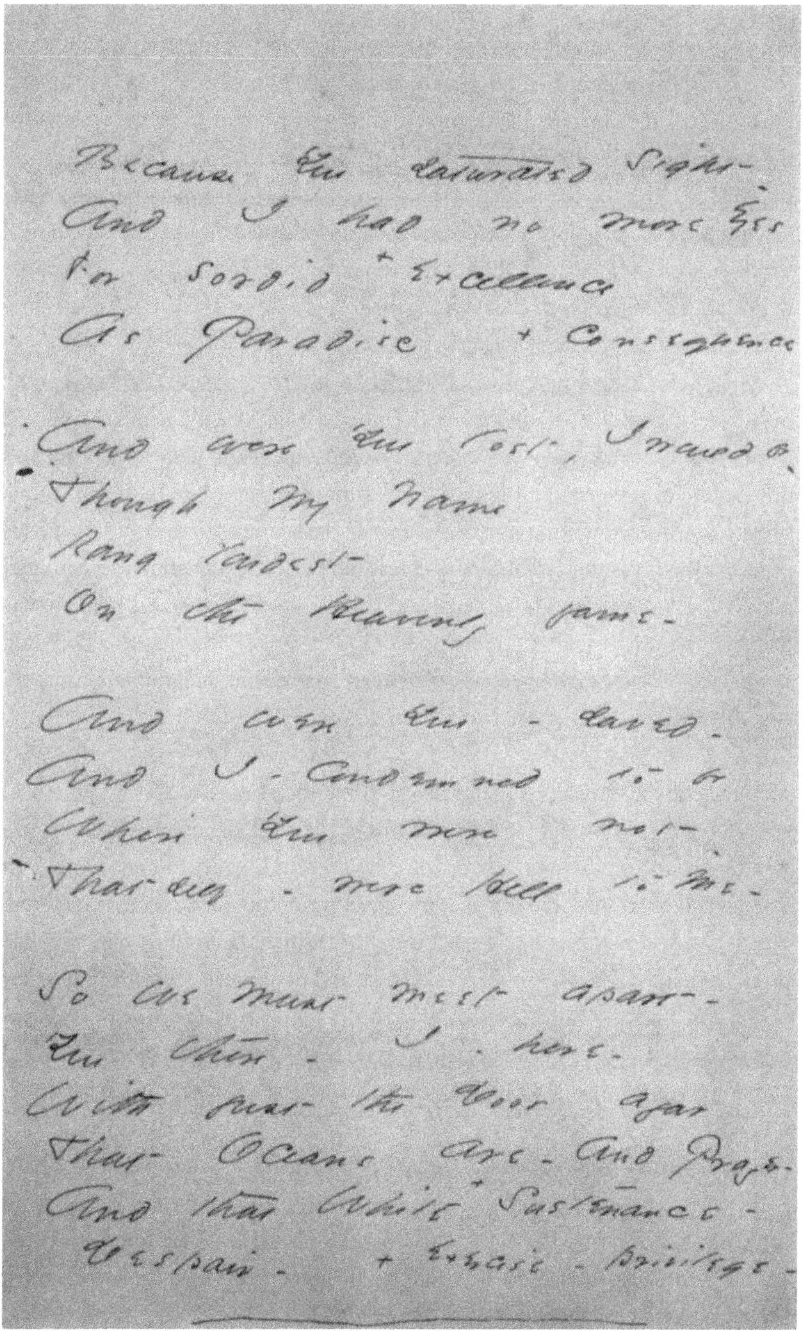

Figure 5 Fascicle copy, Emily Dickinson's "I cannot live with You –". "You there – I – here – " is the second line of the final stanza. From Fascicle 33. Courtesy of Houghton Library, Harvard University.

of even the letter and the phoneme—indeed, below the level of the grapheme. The second half of the line, "– I – here –" might be taken to emerge from the subgraphemic elements of "there." Dickinson's "t" transforms into "I" as the crossbar of the former splits in half to form dashes that both separate and conjoin the vertical stroke of "I" with the remainder of this rupture, "here." This paragrammatic subtext is more legible in the fascicle text of Dickinson's poem. As usual in Dickinson's handwriting, the "t" in "there" is not crossed but is rather formed by the proximity of a vertical stroke (which looks precisely like Dickinson's "I") to a free-floating horizontal stroke on its right (Fig. 5).

In the fascicle text, the "crossbar" of the "t" is already a dash that has literally been placed *over* "there"—before, between, and after which it is then dispersed to reveal the latency of "– I – here –" within the word "there."[16] Dickinson's line begins to resonate ontologically at the intersection of its two constitutive properties: the paragrammatic dispersion of dashes enacts a sifting of shifters in which the intimately entangled relationship between our primary existential placeholders—"you" and "I," "there" and "here"—is made manifest by the invisible, yet potentially legible, rupture of a graphic mark. Grasping this potential significance of the line demands that we read an invisible, subgraphemic dimension of writing operating prior to signification.

This must be the dimension in which Shanxing Wang's ideal reader is immersed, scanning a collection of poetry books in the National Nanofabrication Laboratory under an ultrahigh-magnification AFM, through which he encounters an assemblage of entities that are at once objects and signs, abstract and material. This is the scene of reading and writing from which, beginning with which, I propose that we approach materials science and materialist poetics as *branches of materials research and fabrication*. To situate these at the *limits* of fabrication is to open a space between "there" and "here" in which we are approached by bodies and words, in which the poetic image gives way onto invisible structures, wherein text passes over into texture. "The mode of presentation of a limit in general cannot be the image properly speaking," writes Jean-Luc Nancy. Rather, "the image properly speaking presupposes the limit which presents it or within which it presents itself. But the singular mode of the presentation of a limit is that this limit must be reached, must come to be *touched*: one must change sense, pass from sight to tact."[17] *The Limits of Fabrication* will articulate the poetics of this situation, in which the specular, aesthetic appreciation of "the image" passes into the material problem of formal construction.

INTRODUCTION: MATERIALS SCIENCE, MATERIALIST POETICS

What is poetry? To say that poetry is making, *poiēsis*, is to offer a classical answer, reiterated across centuries. But to interpret this answer is to provoke argument. For example, Heidegger's understanding of poetry as "a kind of building" might seem aligned with an understanding of poetry as making.[1] But for Heidegger *poiēsis* is bringing-forth or revealing; it is a modality of letting-appear or letting-dwell, and as such it is not understood as making, as manufacturing, or construction. For Heidegger, these are inessential definitions of *poiēsis*, inadequate to the vocation of authentic poetic creation.[2] In *The Limits of Fabrication* I pursue the contrary position that poetry, as making, is a practice of material construction. Since this is a specifically *materialist* position, to adopt it is to take sides, though it is not to say anything especially unfamiliar. But to say that poetry is a branch of materials research and fabrication is to make a more unusual claim demanding elaboration. That elaboration is the task of this book. This book returns to a familiar idea: that poetry, *poiēsis*, is a practice of making. It takes a materialist position on this claim by treating poetry as a practice of material construction. And it seeks to make this claim new by situating it in an unfamiliar context: at the limits of fabrication.

As an act of critical defamiliarization, this displacement relies upon two maneuvers. First, I want to test the idea that poetry is form of making, material construction, by taking that idea literally, examining theories and practices of poetic making alongside current technoscientific methods of manufacturing. In what follows, *technics* and *poetics* will be drawn together by addressing both under the more general rubric of *fabrication*. My topic, then, is not properly relations between "science and poetry" or "*technē* and *poiēsis*." Rather, my topic is fabrication. I study poetry and technoscience as two modalities of this encompassing category. Second, I want to update the way we think about poetry and technoscience as practices of material construction by situating these at the *limits* of fabrication. By this I mean to denote, on the one hand, my attention to experimental poetry—specifically, to a tradition of materialist poetics committed to pushing the formal boundaries of poetic making. On the other hand, "the limits of fabrication" is a phrase used to denote the liminal scales of material construction made accessible to materials scientists and engineers by the instrumental and experimental innovations of nanotechnology, which make it possible to characterize matter in a controlled fashion at molecular or atomic scales less than one-billionth of a meter.[3] Insofar as they are situated at "the limits of fabrication," both materials science and materialist poetics engage the boundaries of formal invention and material construction in their respective fields. These fields are, to say the least, markedly different. But they are nevertheless related insofar as their boundary work draws them into a common terrain of formal, constructive, and ideological problems attendant upon experimental practices of making. The speculative gesture of this book is to address these different fields upon this common ground, considering both materials science and materialist poetics as different branches of *materials research and fabrication*.

In *The Limits of Fabrication* I study a vector of postwar and contemporary poetry and poetics broadly situated within what has been called "the Pound tradition."[4] This is work that affirms and carries forward the commitment of the modernist avant-garde not only to inventing new poetic forms but to doing so through engagement with "the materials of poetry."[5] In other words, in the work of the poets I study a commitment to formal experiment is constitutively bound up with a commitment to materialist poetics. The poetry of Charles Olson, Ronald Johnson, Caroline Bergvall, Christian Bök, and Shanxing Wang addresses composition as the arrangement of material units in a spatial field, and it binds the "meaning" of poetic artifacts to the material form of their construction.[6] Such writing is *materialist* insofar as it is *constructivist*, and it situates the practice of poetry within the wider field of material construction I demarcate by the term "fabrication."

An example of this commitment to materialist constructivism is offered by the journal *Facture* (2000–2001), which defines the practice denoted by its title as follows:[7]

> act of making,
> formation 1. The
> manner in which
> something is made or
> finished: Execution:
> often: the quality or
> handling of a surface (as
> in painting). 2. Archaic:
> the act or process of
> making something. 3.
> The manner in which a
> surface is dealt with:
> Fact ~ Manufacture
> Fracture ~ Refraction.

Similarly, the word *fabricate* means "to make anything that requires skill; to construct, manufacture." The term conjoins *poiēsis* (making) and *technē* (skill) in a constructive project of skilled making or technical formation. According to the *OED*, to fabricate is "to form (semi-finished metal stock or other manufacturing material) into the shape required for a finished product." And to fabricate is also "to 'make up'; to frame or invent (a legend, lie, etc.); to forge (a document)." Fabrication is thus a practice of material/discursive construction or artifice: the physical production of a material/textual artifact. What Louis Zukofsky calls "the materials of poetry" are the "other manufacturing material" from which a poem is constructed or formed as an articulated shape or structure. Understood in these terms—as an act of making, formation, construction, manufacturing that is at once material and discursive—the poetry I study pushes the *limits* of fabrication insofar as it experiments with the invention of new poetic forms through an engagement with the fundamental materials of poetic language (mark, space, grapheme, phoneme, breath, sound, signifier). In this vein, Steve McCaffery directs our attention toward the "protosemantic" dimension of poetic language that is "prior to meaning," calling for "a serious consideration of both a residual and a possible micropoiesis."[8] According to McCaffery, the protosemantic is

> more a process than a material thing; a multiplicity of forces which, when brought to bear on texts (or released in them), unleash a combinatory fecundity that includes those semantic jumps that manifest within letter shifts and verbal recombinations, and the presyntactic violations determining a word's position: rupture, reiteration, displacement, reterritorialization.[9]

Exploring the protosemantic dimension of poetic language demands attention not only to poetic form but to the manner in which poetic forms emerge from the constitutive materials of poetic writing. The poetic theories and poetic artifacts produced by the writers I study bind the problem of formal invention to the structural disposition and constituent elements of "the materials of poetry." They take these two problems to be inextricable, viewing the formal disposition of the materials as inherent to their constitution, and vice versa.

If *materialist poetics* engages the fundamental materials of poetic language in order to experiment with the production of new poetic forms, *materials science* studies the fundamental structures of matter in order to fabricate materials whose precisely determined formal constitution endows them with novel physical properties. This is a broad and venerable field of technoscience that includes ceramics and metallurgy, as well as the production of newer materials such as polymers, composites, and biomaterials.[10] Insofar as my focus is the *limits* of fabrication, my attention will be primarily focused upon relatively recent developments in materials science enabled by nanoscience and nanotechnology. These are important developments because nanoscale instrumentation makes possible the characterization of materials with unprecedented precision, and the capacity to characterize the molecular or submolecular structure of materials in a more controlled fashion than allowed by traditional chemistry has opened the possibility of translating novel molecular structures into specific molar properties. As Philip Ball points out in *Designing the Molecular World: Chemistry at the Frontier*, "the explosion of interest in materials science in recent years has gained tremendously from the realization that an understanding of materials at the molecular level can lead to the design of properties at the engineering level."[11] According to IBM scientist Donald Eigler, the "milestone in man's ability to build things" heralded by atomic positioning (discussed in the preface to this book) "is the ability to build things *from the bottom up*, by placing the atoms where we want them. Atomic scale construction embodies the idea that the structure is exact in the sense that (within manufacturing tolerances) we build just exactly what we want and nothing else."[12] Exploring the application of these advances in instrumentation and engineering to materials science, Philip Ball argues in *Made to Measure: New Materials for the 21st Century* that nanoscale fabrication enables the development of materials whose properties "are planned from and built into their atomic structure . . . whose composition and structure are specified at the smallest scales, right down to the atomic, so as to convey properties that are useful."[13]

In materials science, the limits of fabrication are thus demarcated by two criteria: by the capacity to *characterize* materials at molecular and submolecular scales and by the capacity to *translate* the effects of such structural precision into novel macroscale properties. The authors of the textbook *Nanochemistry: A Chemical Approach to Nanomaterials* stress the importance of this latter criterion: "It is no longer the structure and the composition of the material that is new," they write, "but rather its form and scale, [the manner of] guiding its arrangement and integration, that is new and which yields new materials with new function and new utility."[14] In other words, what is crucial to making it new in materials science is not only the *technics* of engineering (the technical capacity to characterize matter at unprecedented scales) but also the *poetics* of material construction, the *way* in which novel structures and compositions are made. At the limits of fabrication, *technē* and *poiēsis*, technique and making, are more difficult to distinguish than ever. As designer Bruce Mau and his collaborators at the Institute Without Boundaries emphasize in *Massive Change*, part of the task of nanoscale materials science is to fabricate "self-organizing" materials that self-assemble from carefully designed constituents. "Materiality has traditionally been something to which design is applied," Mau's team writes, but "new methods in the fields of nanotechnology have rendered material as the object of design development. Instead of designing a thing, we design a designing thing."[15] It is this turn toward the fabrication *of* materials, rather than fabrication *with* materials, that leads Northwestern professor of materials science and engineering Gregory B. Olson to speak of "designing a new material world," or Washington University professor of materials science and engineering Mehmet Sarikaya to declare: "We are on the brink of a materials revolution that will be on par with the Iron Age and the Industrial Revolution. We are leaping forward into a new age of materials" (Figs. 6 and 7).[16]

Such proclamations, often hyperbolic and loaded with unexamined ideological implications, are part of the technocultural context this book will consider. This will be a *critical* investigation, thinking carefully about the ideological implications of concepts like "design" and of the technoscientific desire to fabricate "a new material world." Conceptual distinctions between life and non-living matter, between organic and inorganic form, criteria of structural exactitude and constructive precision, the rhetoric of conquest and of leaps forward: all of these are at issue not only for materials science but also for materialist poetics in the twentieth and twenty-first centuries. Formal efficiency is a constitutive desideratum not only of nanoscale materials science but also of modernist poetics: "use no super-

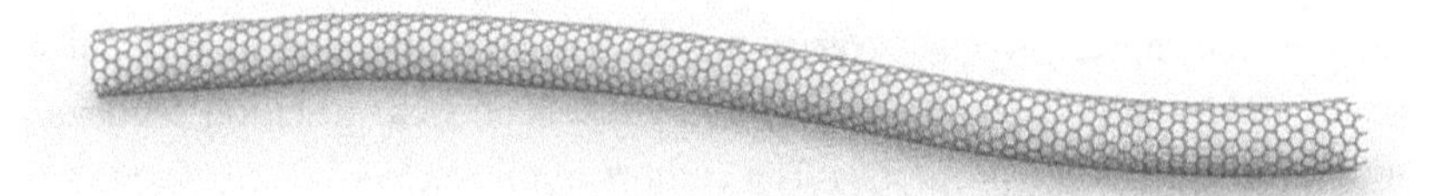

Figure 6 Carbon nanotube. Nanotubes are cylindrically wrapped sheets of hexagonally organized carbon, one atom thick, with a diameter close to one nanometer. They are currently the strongest known material. Courtesy of James Hedberg.

fluous word," counsels Ezra Pound's Imagist doctrine of 1913.[17] Just as Pound's modernist criterion of formal efficiency was in keeping with the formal and aesthetic values of his own machine age, Steve McCaffery suggested in 2001 that "if letters are to words what atoms are to bodies—heterogeneous, deviant, collisional, transmorphic—then we need earnestly to rethink what guarantees stability to verbal signs. . . . In our age of incipient miniaturization, it might be apt to return to the rumble beneath the word."[18] Again, although analogies like McCaffery's Lucretian formula ("letters are to words what atoms are to bodies") will not necessarily be taken at face value in what follows, understanding the context in which such an analogy is deployed in an "age of incipient miniaturization" is crucial to grappling with the manner in which poets have thought of their own work as a practice of material construction and with how they have attempted to incorporate models from science and technology into their materialist practice. McCaffery's formulation situates the relation of materialist poetics to materialist philosophy in the context of contemporary technoscience. Focusing on nanoscale materials science as part of that context, *The Limits of Fabrication* considers how poetics, science, and philosophy all bear upon the present status of *materialism* as an orientation in thought and practice potentially coarticulated by these different fields. Thus, my effort in this book is not only to study models and practices of fabrication in materials science and materialist poetics. More fundamentally, I want to think through the way in which a twenty-first-century materialism might be conditioned by poetry, science, and philosophy.

FROM FIGURATION TO CONFIGURATION

In *Toy Medium: Materialism and the Modern Lyric*, Daniel Tiffany poses the following question to literary critics, historians of science, and philosophers:

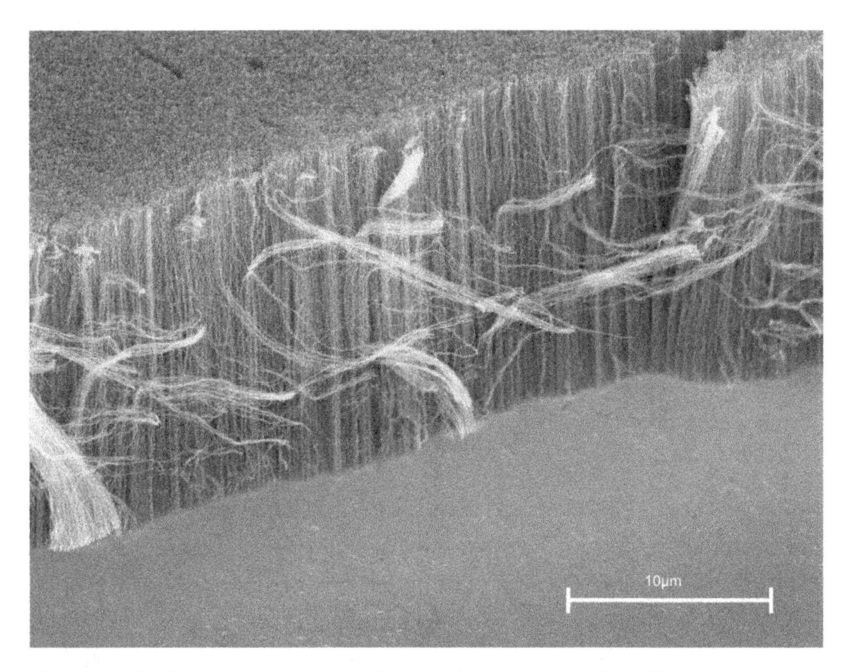

Figure 7 Carbon nanotube array. Nanotubes are grown in self-organizing arrays or "forests." Courtesy of Chris Williamson, University of Cambridge.

"Does poetics have a place in the history of materialism?"[19] By situating this question at the limits of fabrication, I propose a contextual displacement of that question: Does poetics have a place in *contemporary* materialism? What is the place of poetics in the *future* of materialism? How have relatively recent scientific developments—like the application of nanotechnology to materials science—altered the technocultural context in which we think about the relation of poetics to materialism?

Much of Tiffany's fascinating book addresses the history of microscopy and the consequences of its instrumental limitations for scientific epistemology and materialist philosophy. With admirable care and imagination, he studies both the technical processes and the rhetorical strategies through which early modern science and twentieth-century physics have constructed images of material substance. Tiffany's primary concern is the ineradicable invisibility of material substance below certain scales and the paradoxes that ensue from the impossibility of directly producing pictures of so-called primary corpuscles. Since single atoms are smaller than a wavelength of light, and thus below the diffraction limit of optical microscopes, they are inaccessible to visual perception. Yet, according to Tiffany, the empiricist epistemology undergirding scientific materialism demands verification of their existence through direct observation. *Toy Medium* explores the following problem: materialist philosophy and scientific materialism posit the existence of primary corpuscles, or atoms, but since these are invisible their existence remains inaccessible to empirical verification. The materialist cannot see that in which he or she wants to believe. Tiffany thus invites us to consider what he regards as an abiding paradox of materialism:[20]

> The introduction of Epicurean atomism establishes the framework for an irresolvable conflict between ontology and epistemology within atomist doctrine: only atoms truly exist, yet, according to the premises of empiricism, they are unknowable and indeed inconceivable.[21]

In *Toy Medium*, it is the poetic *figuration* of material substance that intervenes in this conflict between materialist ontology and epistemology. Tiffany shows that materialist philosophy and scientific materialism have both relied upon imaginative *figures* of invisible corpuscles (toys and automata, snowflakes and meteors, radioactive bodies and leaping animals) in order to depict or "picture" a material substance we cannot see.

On this account, the place of poetry in the history of materialism hinges upon the function of the poetic image, specifically the capacity of poetry to formulate figurative images of invisible matter. *Toy Medium* argues that "the foundation of material substance is intelligible to us, and therefore appears to be real, only if we credit the imaginary pictures we have composed of it." Since the resolution limits of optical microscopy ensure that these pictures are necessarily "imaginary," the production of poetic images of material substance plays an important role in the rhetorical economy of materialism. "If the substance of reality is the substance of illusion in poetry," Tiffany concludes, "the substance of the lyric therefore remains an essential element in the fabrication of the real."[22] What Tiffany calls "lyric substance" supplements the incapacity of scientific instrumentation to meet the demands of scientific epistemology, and this supplement turns out to be essential insofar as the lyric figuration of matter has always been central to the discourse of scientific and philosophical materialism.

My own inquiry into the role of materials science and materialist poetics in "the fabrication of the real" begins where *Toy Medium* leaves off. Tiffany demarcates the boundaries of his study precisely at the moment in the history of technics when developments in nanoscale microscopy displace the materialist paradox theorized in *Toy Medium*. Tiffany specifically notes that his arguments do not account for the pertinence of non-optical microscopy to the lacunae of observation with which he grapples. "My concern," he writes,

> is not whether atoms actually exist (a fierce point of debate in modern physics), or whether new electronic devices (such as the scanning tunneling microscope) can truly visualize the foundation of material substance. Rather, even if we now possess machines that can take pictures of atoms, the history of philosophical materialism—and its many popular manifestations—is nevertheless conditioned by the invisibility of material substance and by the regime of analogy associated with it.[23]

As we have seen, the Scanning Tunneling Microscope does not exactly "take pictures" of atoms, though it does produce highly mediated visual representations of their position in a manner that circumvents the lacuna of optical observation grounding Tiffany's account. Not only do scanning probe devices produce visual representations of single atoms with a haptic rather than optical interface, the same interface also enables scientists to *position* single atoms. The problem, then, is not so much whether we can "see" single atoms but rather what we are to do with our capacity to position them with precision control. What is at issue is a pragmatic capacity rather than an epistemological aporia.

Similarly, the crucial role played by "quantum tunneling" in the operation of the STM (a feedback loop is established by measuring the flow of electrons between the tip of the microscope and the surface of the sample) exemplifies another way in which accounting for nanoscale instruments tends to displace the epistemological lacuna with which Tiffany is concerned. "At present," he argues, "we have no way of reconciling the nature of quantum materiality (the physics of subatomic matter) with the experience of ordinary bodies."[24] The irreconcilability of quantum mechanics and everyday experience is thus, for Tiffany, another site in which lyric intervenes by figuring quantum phenomena through the resources of the poetic image. But if Tiffany's statement concerning the rift between our knowledge of quantum mechanics and our experience of middle-sized objects seems intuitively true, nanoscale materials science nevertheless promises to draw these scale levels together by sidestepping the epistemological aporia *Toy Medium* foregrounds. The task, in this case, is not the epistemological or phenomenological problem of reconciling quantum mechanics with our intuitive, embodied sense of physical law but rather the practical problem of how to engineer structures such that phenomena particular to nanoscale matter have effects upon the physical properties of macroscale materials. The STM harnesses the phenomenon of quantum tunneling in order to configure nanoscale structures through the mediation of a macroscale device. From this perspective, the discrepancy between quantum and Newtonian mechanics is less an epistemological obstacle than an engineering opportunity:

> In engineering terms, there seems so far to be nothing we can do on the macroscale that we can't do at the nanoscale. . . . That's not to say that nanotechnology is no more than macrotechnology writ small. Some things are very different down at the scale of molecules and cells. Fluids start to look grainy and highly viscous. Surface tension becomes a dominant force. Quantum effects come into play, and everything is buffeted by Brownian motion. But with a bit of ingenuity, all of these things can be used to advantage.[25]

The assertion that "with a bit of ingenuity" quantum phenomena can be used to advantage by engineers implies that, while the formal and phenomenal differences between quantum and Newtonian mechanics remain a theoretical question for the epistemology of science, refined *practices* of material construction have begun to circumvent that problem by fabricating devices and materials that harness quantum effects in order to alter macroscale properties. In this sense, what was once an epistemological quandary might now more pertinently be considered an engineering problem. In order to move from *Toy Medium*'s concern with the history of microscopy and the history of philosophical materialism toward the contemporary concerns of materials research and fabrication, we need to understand how the capacity of new instruments to both image and manipulate nanoscale matter alters the imaginary economy of materialism at issue between science, poetry, and philosophy. My argument is that nanotechnology shifts the ground of "the fabrication of the real" from *figuration* to *configuration*.

Scanning probe devices exemplify this shift insofar as they displace the epistemological lacunae of scientific objectivity by foregrounding the pragmatics of material construction. In *Objectivity*, their important study of the history of that concept in scientific epistemology, Lorraine Daston and Peter Galison note that by the early twenty-first century,

> visual presentation was becoming part and parcel of the making of new kinds of things, from quantum dots to switchable nanotubes. It is no accident that even the first generation of university nanolaboratories integrated visualization facilities architecturally within the fabrication facility. It is frequently not possible to make things without depicting them visually—and, quite often, it is not possible to represent them without the procedure of making. The atomic force microscope and the scanning tunneling microscope were perfect examples of this compound: the same device was used at the same time to image and to alter.[26]

Nanoscale fabrication fuses mimesis and making. As Colin Milburn argues in *Nanovision: Engineering the Future*, "Nanotechnology marks a point in scientific history where representing and intervening become one."[27] In these terms, Tiffany's promotion of the poetic image as "an indispensable guide to reality" is predicated upon the incapacity of technoscience to represent, objectively, the objects upon which he claims empirical epistemology and materialist ontology depend. But Daston and Galison's account makes clear the contemporary limitations of this approach: "In the realm of nanomanipulation images are examples of right depiction—but of objects that are made, not found. . . . In this corner of science, the representation of the real—the use of images to finally get nature straight—may be coming to a close."[28] At the limits of fabrication, technoscience circumvents the limits of objectivity by *configuring* invisible worlds even as it renders visible their mediations.

In this technocultural context, poetry may have to surrender its hallowed vocation as primordial guardian of the image, acknowledging the diminished power of its privileged capacity to present the unpresentable by producing *figures* of invisible worlds. To continue, today, to pursue the proper locus of poetry's participation in the economy of materialism via the figurative power of the poetic image is to chase a red herring. Rather, the locus at which we might think the relation of science and poetry vis-à-vis materialism shifts from figuration to configuration, from fabulation to fabrication, from the evocation of images to the making of material structures, from *mimesis* to *poiēsis*. This is the context in which we might return to and reexamine the ancient determination of *poiēsis* as making, of poetry as a form of building or a practice of material construction, and in which we might defamiliarize and radicalize these determinations of poetic practice by considering the latter as a branch of materials research and fabrication.

While Tiffany's *Toy Medium* builds a compelling argument for the materiality of the poetic image as "lyric substance," he largely leaves aside the problem of poetic form and the poetics of material configuration, paying little attention to the scriptural marks and formal assemblages that are the *material precondition* for articulating images accessible to imagination and interpretation. Accordingly, his readings of poetic examples are primarily thematic, attending to figurative language and explicating its meaning. But if meaning is the vehicle of lyric substance, it remains to consider the physical configuration of materials undergirding the construction of poetic meaning. Consider the difference, for example, between Emily Dickinson's line "You there – I – here –" and Jorie Graham's lines from "Pietà":

Listen. Do you hear it at
last, the spirit of
matter, there, where the words end—their small heat—where the details
cease, the scene dissolves, do you feel it at last, the sinking, where the
meaning
rises, where the meaning evaporates[29]

Graham presents us with the mise-en-scène of a hermeneutic and epistemological aporia, explicitly evoking an invisible absence at the limits of language as "the spirit of matter." Dickinson's line does not describe such an absence; it does not mention matter; it is not "about" the dissolution of the scene of writing. Rather, whether intentionally or not, the line enacts such a dissolution, makes it *happen*; it constructs this dissolution, as an invisible material event that is nonetheless legible in the material configuration of an inscription.[30] "Prior to meaning," in Steve McCaffery's terms, we do not arrive at the place "where the / meaning / rises, where the meaning evaporates" by contemplating the poet's suggestion that there may be such a locale, nor through an affective swoon induced by the profundity of an evocative question. In Dickinson's line, that place arrives at us if and when we submit the material particulars of the line to close inspection. The "meaning" does not rise, but the place where it evaporates does—without depending upon meaning to do so. Without semantic permission.

SIGNALETIC MATERIAL

Thus, while *Toy Medium* studies poetic images of material substance (weather in Wallace Stevens, toy birds in W. B. Yeats, the jerboa's leap in Marianne Moore), attention to the protosemantic dimension of poetic language will involve studying the concrete fabrication of poetic forms with the materials of poetry. This will involve turning to another poetic tradition. We might grapple with the manner in which both the "hybrid materials" engineered by nanotechnologists and the "objectist" poetics of Charles Olson challenge us to displace models of organic form predicated upon essential distinctions between living and nonliving matter. We might study the influence of Buckminster Fuller's geodesic architecture upon the form of Ronald Johnson's long "architectural" poem *ARK*, as well as Fuller's influence upon the chemists who discovered the structure of a carbon molecule they named the Buckminsterfullerene, which has become a cornerstone of nanoscale materials science. We might examine crystallography and molecular biology as fields that afford structural models not only for the development of self-organizing, nanostructured biomimetic materials but also for the materialist poetry of Christian Bök (*Crystallography*) and Caroline Bergvall (*Goan Atom*). Through Shanxing Wang's *Mad Science in Imperial City*, we might gauge the degree to which it is mathematical formalization, rather than figurative images, that mediates the relation of scientific materialism to its invisible objects, considering the consequences of that recognition for materialist poetics. And from a historical materialist perspective, we might situate the 'pataphysical historiography of Wang's book by thinking through the contemporary political and economic conditions of its double demand to "work nano, think cosmologic."[31] These are some of the tasks taken up by the chapters that follow.

To clarify the methodological stakes of such an approach, let me turn from *Toy Medium* to another work addressed to materialist criticism, in this case in the mode of critique rather than elaboration: Walter Benn Michaels's *The Shape of the Signifier*. Michaels objects to "the replacement of the sign by the mark"[32] in materialist criticism and the latter's consequent attention to "material events,"[33] a tendency that he traces to Jacques Derrida's and Paul de Man's insistence upon "the materiality of the signifier" and which he finds most insidiously operative in Susan Howe's work on the manuscripts of Emily Dickinson.[34] According to Michaels, an appeal to the materiality of the text will always rely upon an appeal to the experience of the reader, which will in turn devolve into a relativist appeal to a subject position or identity. "I am arguing," he writes,

> that anyone who thinks that the text consists of its physical features (of what Derrida calls its marks) will be required also to think that the meaning of the text is crucially determined by the experience of its readers, and so the question of who the reader is—and the commitment to the primacy of identity as such—is built into the commitment to the materiality of the signifier.[35]

For Michaels, the problem with this purported "commitment to the primacy of identity" is that it covertly shifts our attention from the *object* to the *subject*: "the question of what's in the work of art (a question about the object) is replaced by the question of what the reader sees (a question about the subject)." "Because what something looks like must be what it looks like to someone," Michaels argues, "the appeal to the shape of the signifier is at the same time an appeal to the position and hence to the identity of its interpreter."[36] On these grounds, Michaels urges a return to authorial intention as the locus of meaning, offering us the following alternative:

> The argument, in miniature, is that if you think the intention of the author is what counts, then you don't think the subject position of the reader matters, but if you don't think the intention of the author is what counts, then the subject position of the reader will be the only thing that matters.[37]

Either we refer to the intention of the author in order to answer a question about the object ("what's in the work of art"), or we impose our own "identity" upon the text by referring to the phenomenal experience of a subject ("what the reader sees").

Amid the bevy of false oppositions and specious syllogisms from which his critique of materialist criticism is constructed, the main problem with Michaels's appeal to authorial intention is that such an appeal in no way anchors interpretation in the textual "object"; rather, it merely shifts the locus of authority to another "subject." Moreover, Michaels fails to account for the fact that any critical judgment concerning authorial intention will itself require an act of "subjective" interpretation on the part of a reader that, at least by Michaels's lights, would devolve back onto the identity of the interpreter. That is: "if you think the intention of the author is what counts," then "the subject position of the reader" will still matter because it will potentially affect a judgment concerning authorial intention just as much as a judgment concerning the shape of the signifier. Thus appeals to authorial intention cannot suffice to obviate, on their own, the kind of relativism that Michaels imputes to materialist criticism, since interpretive recourse to authorial intention still involves particular judgments concerning authorial intention and its interpretive significance.

Yet even if we grant Michaels's recourse to authorial intention, in many cases the latter will support the protocols of the materialist critic. Indeed, I agree with Gilles Deleuze's remark in his book on Francis Bacon that "we do not listen closely enough to what painters have to say,"[38] and the same applies to poets. In *The Limits of Fabrication*, part of the reason I attend to the physical features of the text is that this is precisely what the poets I study ask of their readers. These authors are eager to dissolve their putative authority into what has been called "the materiality of the signifier" or what Zukofsky calls "the materials of poetry." Charles Olson elaborates a poetics and a practice of writing attentive to the *physicality* of both the body and the text at work in processes of composition and reading that subsume the dualism of subject and object. Caroline Bergvall describes her work's address to genetic engineering in terms of a recombinatory engagement with the material organization of a textual body. Christian Bök reissued his book *Crystallography* in 2003 because the typesetting of the original

1994 edition was insufficiently precise to convey "the shape of the signifier" and therefore the sense of the poem. Shanxing Wang describes his handling of pronouns in terms of "a 'blurring' of the object, of the material object with the symbol,"[39] and he does not hesitate to elaborate his point by way of an analogy between pronouns and the symbols of the periodic table. Ronald Johnson describes *ARK* as "literally an architecture . . . fitted together with shards of language, in a kind of cement music."[40] Steve McCaffery tells us, "I try to realize a materialist poetics of formlessness," and he asks that we consider both writing and reading as a "material scene of forces"[41] agitated by protosemantic energies latent in both inscription and interpretation. Thus Michaels's defense of authorial intention against materialist criticism collapses as soon as we take seriously the authorial intentions at play in a whole tradition of materialist poetry and poetics. We might also note that this is precisely the tradition in which the critical and poetic work of Susan Howe participates.

Consider a page from the second panel of Steve McCaffery's *Carnival* (1967–75), a "multi-panel language environment"[42] intended to be torn out, page by page, from the offset book in which it was printed and assembled as a sixteen-sheet compound wall panel. McCaffery characterizes *Carnival* as "language presented as direct physical impact." "The thrust is geomantic," he writes, "realignment of speech, like earth, for purposes of intelligible access to its neglected qualities of immanence and non-reference."[43] The reader is called upon to function as a "perceptual participant" in the construction of the work, for the good reason that the "defective messages" it relays are refractory to linear reading.[44] The foregrounding of "the shape of the signifier" in such a text—a nonlinear distribution of material inscriptions demanding visual as well as literary interpretation—activates language as a spatial construction that is both seen and read: looked at, interpreted, traversed. The physicality of the text engages the embodied cognition of the reader in an effort to construct the legibility of inscriptions: to see them, make them out, while the intention of the author and the objectivity of the text enter into this construction insofar as they at once elicit and obstruct it (Fig. 8).

We might consider a text like *Carnival* in terms of what Gilles Deleuze, in *Cinema 2*, calls "signaletic material." Taking us beyond the restricted purview of "the materiality of the signifier," signaletic material can include "all kinds of modulation features, sensory (visual and sound), kinetic, intensive, affective, rhythmic, tonal, and even verbal (oral and written)."[45] Though Deleuze is discussing the cinematic image, the sort of verbal material he includes in this catalog might include such works as *Carnival*, which engage a protosemantic dimension of prelinguistic material. Though "prior to meaning," these materials are constantly on the cusp of semantic transformation. Signaletic material, Deleuze writes,

> is a plastic mass, an a-signifying and a-syntaxic material, a material not formed linguistically even though it is not amorphous and is formed semiotically, aesthetically and pragmatically. It is a condition, anterior by right to what it conditions. It is not an enunciation, and these are not utterances. It is an *utterable*. We mean that, when language gets hold of this material (and it necessarily does so) then it gives rise to utterances which come to dominate or even replace the images and signs, and which refer in turn to pertinent features of the language system, syntagms and paradigms, completely different from those we started with. We therefore have to define, not semiology, but "semiotics", as the system of images and signs independent of language in general. When we recall that linguistics is only part of semiotics, we no longer mean, as for semiology, that there are languages without a language system, but that the language system only exists in reaction to a *non-language material* that it transforms.[46]

The problem with approaching materialist poetics or materialist criticism under the banner of "the materiality of the signifier" or "the shape of the signifier" or even "the materials of poetry" is that these genitive constructions—which address the materiality *of* the language system—can occlude attention to the manner in which the materiality of language is predicated upon the transformation of "non-language material." It is not only the "letters on the page" or the "shape" of the signifier that is foregrounded by modernist traditions of sound, visual, and concrete poetry running from Filippo Marinetti and Hugo Ball through Caroline Bergvall; more primarily, these traditions foreground the *construction* of poetic forms and utterances from protosemantic materials. That is to say, it is the material *fabrication* of poetry—*poiēsis* as a practice of material construction—to which we have to attend if we hope to account for the manner in which authorial intention, textual marks, and readerly agency collide in such works as McCaffery's *Carnival* or Ronald Johnson's *ARK*.

The atomic IBM logo fabricated at Almaden Labs in 1989 exemplifies the transformation of non-language material by a language system, the transformation of the protosemantic into the semantic, of "signaletic material" into linguistic significance, of materials into the materiality of the signifier. It is not only the "shape" of the signifier that is at issue here but also the *way* in which a material form is made, constructed. The atomic IBM logo not only signifies the

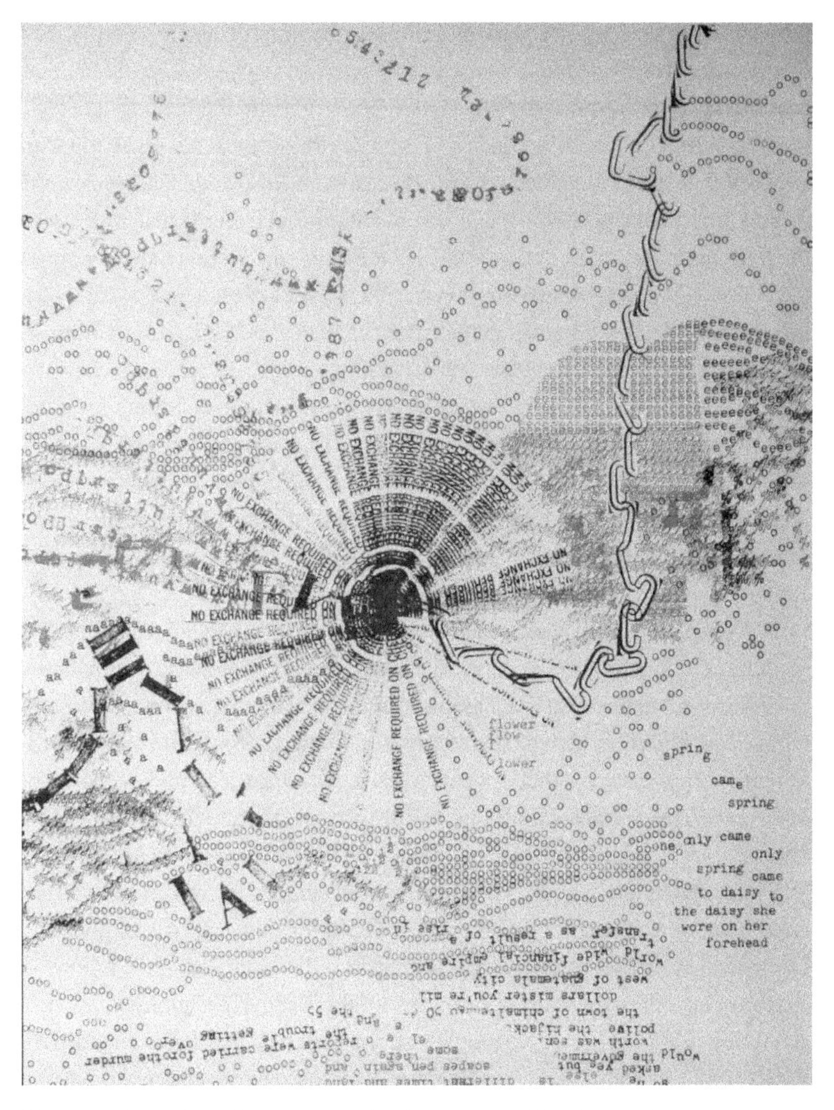

Figure 8 Page from Steve McCaffery's *Carnival, The Second Panel: 1970–1975*. Courtesy of Steve McCaffery.

corporate acronym it spells out through the formal distribution of its components; it signifies a specific *practice* of making, a new way of building material forms. It advertises this as a capacity of the corporation the sign represents. And this practice, resolved into a material signifier, also implies a specific manner of ***thinking*** about form, about the construction of form: the technics (*technē*) of its construction entails a "poetics," if we take this in the strict sense to mean a theory of *poiēsis*, of making, of forming materials. Insofar as it advertises the inception of atomic positioning as a potential means of building materials "from the bottom up" or, more dramatically, of "remaking the world atom by atom,"[47] Eigler and Schweizer's demonstration connotes a host of epistemological, ontological, and ideological entailments resonant with the history of atomism in philosophy and technoscientific representation. It also entails a mechanistic approach to manufacturing[48] to which a contrasting method of bottom-up materials engineering, based on chemical self-assembly, has been opposed. Championed in particular by Harvard chemist George M. Whitesides,[49] a self-assembly approach to nanoscale construction relies upon the capacity of nanoparticles to self-organize into coherent structures through chemical interaction. Engineers attempt to guide and specify the integration of these arrangements across scales so as to exploit, within bulk ensembles, particular properties enabled by the specification of nanoscale components. A self-assembly approach to "the limits of fabrication," rather than one based on atomic positioning, involves engineering the *conditions* under which novel materials emerge from processes of autoformation. It utilizes "synthetic chemistry to make nanoscale building blocks of different size and shape, composition and surface structure, charge and functionality" and then arranges the conditions under which these building blocks "organize spontaneously into unprecedented structures, which can serve as tailored functional materials."[50] This approach to nanoscale materials engineering has yielded some of the most significant results to be discussed in the following chapters, including the fabrication of materials adaptively responsive to their environment (Chapter 1), exceptionally strong materials like carbon nanotubes (Chapter 3), and "biomimetic" materials whose construction relies upon the imitation of biological processes and structures (Chapter 4).

As we shall see, concepts of form and models of material construction emerging from materials science also play an important role in shaping the productions of materialist poetics. Here, too, the confluence and conflict of mechanistic atomism, self-assembly, biomimesis, and mathematical formalism are at work in the determination of formal methods, constraints, and procedures guiding the fabrication of poetic forms and structures. This is not an idle analogy, by which "science" and "poetry" are yoked together for comparison by the facile observation that each in some sense involves thinking about form and building material structures. Rather, what is at stake is understanding the way in which concepts and practices of fabrication, as the common ground of materials science and materialist poetics, are technoculturally constructed. Books like *ARK*, *Crystallography*, *Goan Atom*, and *Mad Science in Imperial City* cannot be understood—we cannot begin to work with them as critics—without investigating the technoscientific methods and models of fabrication with which they are engaged at the level of both content and form. Likewise, in seeking to understand the historical and contemporary significance of science and technology, we cannot grapple with the models of form and the methods of making at issue in their development without grasping their conceptual, ideological, and aesthetic implications. These implications are precisely what the works of poetry named above address and with which they experiment, through their discursive and formal engagement with technoscientific models and methods. "Fabrication" thus names one point of entry into the contemporary *episteme* traversing, articulating, and coproduced by the overlapping boundaries of *technē* and *poiēsis*.

METHOD AND STRUCTURE

The methodology of my study is not homogeneous, nor is my attention to the fields I investigate distributed equally across its chapters. I have proceeded according to the principle that each chapter, insofar as it studies a specific problem and a specific set of materials, deserves a specific methodological approach. I should also stress that while the sequence of the chapters arranges the conceptual issues at stake in a particular order, the structure of the book does not necessarily reflect the order of its genesis or the priority of the problems and fields it takes up. For example, the book leads with a chapter on science and philosophy and then carries the conceptual problems raised there into the field of poetics in Chapter 2. But an immersion in Charles Olson's poetics already informs my approach to the problems addressed in the first chapter. Similarly, if a chapter opens with a discussion of science and technology (as in the case of Chapter 4), or if it begins with questions and quotations drawn from poetry (as in the case of Chapter 5), the contingency of beginnings should not necessarily be understood to lend priority—in terms of intellectual authority—to either field. The "interdisciplinarity" of the project demands a synthetic approach throughout, on the part of the reader as much as the writer, and this sometimes involves thinking retroactively or discontinuously, rather than understanding the sequential distribution of the various materials as indicative of relative importance.

The first two chapters should be considered as a single unit, and what draws them together is their philosophical approach to technoscience and poetics. Both address the relation of "the open" to what Heidegger calls "physical being," doing so through an investigation of the ontological and poetic stakes of conceptual distinctions between living and non-living beings. Chapter 1 takes up this problem in relation to conceptual challenges posed by nanoscale materials engineering, Chapter 2 in terms of Charles Olson's materialist poetics. Together, these two chapters form an introductory section attempting to clarify the stakes of a problem at issue throughout *The Limits of Fabrication*: that of how traditional distinctions between living and non-living matter, organic and inorganic form, are troubled by the approaches to materials research and fabrication that I study in both science and poetry. I challenge particular ways of understanding distinctions between living and non-living beings, as well as the idealist ideology of "organic form," and I rethink the relation between concepts of "the object" and "the open" in a manner responsive to recent developments in materials science and to Olson's "objectist" poetics. If the reader should ask "what does this have to do with poetry" while reading Chapter 1, or "what does this have to do with materials science" while reading Chapter 2, I have to beg the reader's patience until Chapters 3, 4, and 5, which take a synthetic approach to the relation between these fields—an approach that is grounded by the philosophical engagement with each field carried out in the first two chapters.

Chapter 1, "The Inorganic Open," examines the implications of "smart materials"—materials responsive and adaptive to their environment—for the distinction between modalities of being proper to the object, to life, and to *Dasein* (stone, animal, and man) developed by Heidegger in his 1929–30 seminar, *The Fundamental Concepts of Metaphysics*. On this basis, I develop a critique of "biopolitics" as an adequate conceptual framework within which to address certain challenges to the concept of "life" posed by contemporary technoscience. Specifically, I criticize the biocentrism of Giorgio Agamben's approach to the concept of "the open,"[51] and I interrogate the relation of that concept to the status of what Heidegger terms "physical being."

In Chapter 2, "Objectism," I take up the conceptual relation between "physical being" and "the open" in terms of Charles Olson's materialist poetics of "objectism" and "open field" poetry. Here I attend to the influence of Alfred North Whitehead upon Olson's poetics, attempting to recontextualize the ontological implications of Olson's work with regard to the present technocultural conjuncture. To that end, I reconstruct the background of Olson's poetics in Riemannian geometry, relativity theory, and quantum mechanics, arguing that his efforts to develop a materialist poetics responsive to post-classical science anticipates the shift at stake in nanoscale materials science: a shift from "what we know" to "what we do," from the epistemological problems generated by post-classical science to pragmatic engagements with their implications for the making of material structures. Engaging other critics of Olson's work, including Marjorie Perloff, Charles Altieri, and Don Byrd, I also try to rethink Olson's specific contribution to a tradition of modernist poetics running from Pound's *Cantos* through the objectivist poetics of Louis Zukofsky and George Oppen to the open field poetry of Robert Duncan, Denise Levertov, and Robert Creeley, distinguishing the import of Olson's "objectism" within this tradition and schematically articulating its connection to some recent projects in experimental poetry. My argument is that, far from endorsing a neo-romantic poetics of "the natural look" or "organic form" (as Marjorie Perloff and Steve McCaffery have argued), Olson's midcentury poetics is, on the contrary, an effort to displace the opposition of organic and inorganic form through a Whiteheadian engagement with the relation between "the object" and "the open." Part of the literary historical upshot of this argument is to align Olson's objectism with a tradition of contemporary poets working at the intersection of "open field" and "procedural" poetics, rather than perceiving these approaches as incompatible on the supposed basis of their discrepant organicist or mathematical inclinations.

If the first two chapters treat a single conceptual problem (the status of "physical being" and "the open" with respect to distinctions between the living and non-living, the organic and inorganic) while holding materials science and materialist poetics apart, Chapter 3, "Design Science," begins to draw these fields together. Here I study the influence of Buckminster Fuller's "design science" upon models of form and techniques of self-assembly in both nanoscale materials engineering and in Ronald Johnson's long poem, *ARK* (1970–91). Fuller was a colleague of Charles Olson at Black Mountain College in the late 1940s, where he worked closely with many of the same students. Thus my investigation of the influence of "design science" upon materials science and materialist poetics enables us to return to the context in which Olson's "objectism" was developed while studying the legacy of Black Mountain College from another angle, sorting out some of the discrepant ways of thinking about form and fabrication which that midcentury milieu continues to transmit. Fuller's geodesic architecture, early prototypes of which were developed at Black Mountain, was a crucial inspiration in the modeling of the C60 molecule—subsequently named the Buckminsterfullerene—in 1985. This breakthrough opened a new approach to nanoscale carbon chemistry and to materials self-assembly, eventually resulting in the fabrication of carbon nanotubes in 1991. Fuller was also a friend of Ronald Johnson (as well as of Ezra Pound, Hugh Kenner, John Cage, and many other poets), and his "design science" was an important influence upon the formal strategies and mimetic priorities of Johnson's work. My goal in this chapter is not only to track the import of Fuller's work as a precedent to structural principles in materials science and materialist poetics but also to discern the ideological consequences, in both of these fields, of approaching material form in terms of "design." In this vein, I critique the ideological complicity of "design" with an idealist concept of "nature." The chapter includes a consideration of the relevance of Emersonian transcendentalism to Fuller's design science and to Johnson's poetics, and it concludes by offering a general definition of the "nature poem."

Chapter 4, "Surrational Solids, Surrealist Liquids," studies the way in which crystallography and molecular biology inform mimetic practices of fabrication in materials science and materialist poetics. Turning to Christian Bök's volume of concrete/conceptual poetry, *Crystallography*, and to Caroline Bergvall's recombinant feminist/queer poetics in *Goan Atom*, I examine the way in which these books deploy and *détourn* structural models drawn from crystal geometry and biotechnology. I situate the mimetic strategies of their materialist poetics in relation to those of materials scientists for whom crystallography provides models of structural self-organization and formal efficiency, or who seek to develop "bionanotechnology" by building "machines, materials, and devices with the ultimate finesse that life has always used."[52] I read *Crystallography* and *Goan Atom* together as the poems of a technoscientific climate in which the "dry" and "wet" worlds of solid-state chemistry and biology are increasingly fused in pursuit of hybrid, "biomimetic" materials. While Bök's and Bergvall's work has been situated in the procedural, constraint-based tradition of the OuLiPo by Marjorie Perloff,[53] I expand the parameters of that lineage and complicate its allegiances by reading their work as a renovation of Olson's objectism, thus situating them in relation to "the Olson tradition" of open field poetry as well as the proceduralism of the OuLiPo. I argue that despite their very different priorities and formal methods, what Bök's *Crystallography* and Bergvall's *Goan Atom* have in common is a contemporary reactivation of an objectist poetics that traverses the organic and inorganic worlds of biology and solid-state chemistry.

Finally, Chapter 5, "The Scale of a Wound," turns to Shanxing Wang's extraordinary 2005 volume, *Mad Science in Imperial City*. Formerly a mechanical engineer involved in nanoscale research toward the end of his scientific career, in 2002 Wang gave up his faculty position in the Department of Engineering at Rutgers in order to focus on writing. The result, *Mad Science in Imperial City*, fuses his knowledge of materials engineering, nanotechnology, mathematics, and cosmology with a fractured autobiographical and historical narrative, including his past as a political activist in 1980s China and his later situation as a writer living in post-9/11 New York. Wang's work addresses the political and ideological implications of nanoscale materials science, critiquing the implicit totalitarianism of its quest for absolute control over the structure of matter. *Mad Science* foregrounds what Wang calls "the politics of error,"[54] an insistence upon the irremediable contingency of matter and of history. But Wang's work also develops a highly abstract formal language, exhibiting a 'pataphysical impulse toward the mathematical formalization of historical processes and the algebraic encoding of subjective experience. The equations and the scientific diagrams constellating the book's pages map a highly complex non-territory opening between two discrepant upheavals of the contemporary East and West (the Tiananmen Square massacre of June 4, 1989, and the World Trade Center attacks of September 11, 2001) while also serving to index the contingent encounters and emotional investments of a life that ricochets between these historical brackets. The importance of Wang's investment in mathematical formalism might be grasped by juxtaposing it against Tiffany's account of the relation between poetic and scientific materialism in *Toy Medium*. While Tiffany argues that the figurative image mediates our relation to imperceptible primary corpuscles, for Wang the formal language of mathematics plays a crucial mediating role between physical theory and the imperceptible materials. Wang's writing takes up mathematical inscriptions as poetic materials, exploring their non-figurative capacity to formalize not only physical phenomena but also the *feeling* of history. My reading of Wang's work attempts to periodize his poetics by filtering it through the historical significance of a single phrase, "work nano, think cosmologic" (an alteration of "think globally, act locally"), drawing out the implications of this imperative through an examination of the complicity of work and thought, the importance of intellectual labor, in Wang's late capitalist career as both an engineer and a poet.

Throughout these chapters, my goal in *The Limits of Fabrication* is threefold. First, I hope to introduce my readers—particularly those trained in literature, media studies, and philosophy—to the significance of a little-studied field of technoscience: materials research and fabrication. I hope to suggest the importance of grappling with materials science for anyone interested in approaches to structure, form, and fabrication. Second, I want to intervene in narratives of postwar and contemporary literary history by tracing a tradition of constructivist, nonorganic poetry and poetics from the mid-twentieth century to the early twenty-first century. In this case, I hope to demonstrate that the texts I study—from Olson's *Maximus Poems* through Johnson's *ARK*, Bök's *Crystallography*, Bergvall's *Goan Atom*, and Wang's *Mad Science in Imperial City*—traverse the opposition of open field poetics and procedural poetics, the first supposedly modeled on the coherence of organic forms and the expressivity of the body, the second based on the application of mathematical or rule-based formal systems. I hope my consideration of such poets as Olson and Bök or Johnson and Bergvall in a common context demonstrates that if these traditions are hardly identical, they are by no means incompatible. "Objectism," in my account, provides a rubric under which to consider these traditions together. Third and finally, at the intersection of materials science and materialist poetics, I want to think through the import of these fields and their practices of fabrication for the elaboration of a twenty-first-century materialism: an approach to materialist thought and criticism attentive to scientific and artistic practices of *making*. Attentive, that is, to developments at the intersection of *technē* and *poiēsis*. Here my principle is that *the modality of being of "matter" is contingent upon its configuration* and that "materialism" must therefore include an understanding of the methods, technologies, and procedures by which materials are configured. The burden of the chapters that follow is to study some of these methods, technologies, and procedures, and to grapple with their aesthetic, conceptual, and ideological stakes.

1

THE INORGANIC OPEN: NANOTECHNOLOGY AND PHYSICAL BEING

A stone cannot behave in this way.

—Martin Heidegger, *The Fundamental Concepts of Metaphysics*

The "open" is neither the vague quality of an indeterminate yawning nor that of a halo of sentimental generosity. Tightly woven and narrowly articulated, it constitutes the structure of sense qua sense of the world.

—Jean-Luc Nancy, *The Sense of the World*

One of the central conceptual issues attending the development of nanoscale materials science is its relationship to distinctions between living and non-living matter—or, to state the problem more pointedly, distinctions between life and matter. Eugene Thacker delineates this issue as follows:

> Nanotechnology—the control of matter at the atomic level—is, like biotechnology, a research field that often promises a disease-free, biologically upgraded, posthuman future. But it is also a set of practices that may transform our notions of what it means to have a body, and to be a body. Nanotechnology works toward the general capability for molecular engineering to structure matter atom by atom, a "new industrial revolution," a view of technology that is highly specific and combinatoric, down to the atomic scale. But, in this process, nanotechnology also perturbs the seemingly self-evident boundary between the living and the non-living (or between organic and nonorganic matter) through its radical reductionism.[1]

By "radical reductionism," Thacker means that "for nanotechnology, all objects—including the body—are simply arrangements of atoms."[2] We might expand upon

Thacker's characterization by saying that nanoscale materials science addresses bodies in general as particular arrangements of materials, resolvable and manipulable at the molecular or atomic scale, whose properties and capacities depend upon the configuration or reconfiguration of those materials. Thus new material capacities can be brought about by new material configurations. The capacity to produce such new material configurations is precisely what is afforded by the refined precision of nanoscale instruments.

Thacker's argument suggests that this way of viewing the relation between material configurations and material capacities, mediated by technology, bears not only upon the properties of matter but also upon modalities of being: upon "the boundary between the living and the non-living (or between organic and nonorganic matter)." This is a commonplace of commentaries upon the conceptual significance of nanotechnology. Nanotech prognosticator Charles Ostman, for example, claims that "the very definition of life" is about to undergo a transformation:

> what is coming are artificially engineered organelles and cells, technologies which combine organic and inorganic materials and substrates into integrated nanoscale systems, biomolecular prosthetics, and intra-cellular modification strategies which will redefine the very essence of what is commonly referred to as life.[3]

If the specifically cellular composition of organic bodies is a potential riposte to nanotechnology's "radical reductionism"—the treatment of all bodies as configurations of discrete atomic parts—Ostman's invocation of "artificially engineered organelles and cells" points to the compatibility of specifically "organic" arrangements of materials with the configuration of inorganic components. For Ostman, this suggests a redefinition of "the very essence of what is commonly referred to as life."

In *Nanovision: Engineering the Future*, Colin Milburn thus argues that speculative discourses surrounding nanotechnology perform a chiasmatic entangling of living and non-living bodies, positioning objects as living bodies while opening up living bodies so as to reabsorb subcellular molecular or atomic materials into "the domain of their becoming-entity."[4] "Entity" thus becomes the term positioned beyond the distinction between living and non-living beings. But the question remains: how are we to characterize the modality of being of these "entities" without falling back upon question-begging distinctions between living and non-living beings?

I want to approach this problem through a rethinking of Giorgio Agamben's engagement with the concept of "the open." In *The Open: Man and Ani-*

mal, Agamben offers a critical examination of Martin Heidegger's distinction between modalities of being proper to animal and man in his 1929–30 seminar, *The Fundamental Concepts of Metaphysics*. In that seminar, Heidegger distinguishes between the *physical being* of the object, the *life* of the animal, and the *Dasein* of man. Focusing his critical attention upon Heidegger's distinction between these last two modalities of being, and thus developing his deconstruction of "the anthropological machine" on this basis, Agamben largely ignores the distinction between "physical being" and "life" upon which, I will argue, Heidegger's distinction between animal and man is grounded. If Agamben thus elides the pertinence of physical being to the concept of the open, I am interested in reconsidering that concept in light of this omission. The point of doing so is twofold. First, to develop an approach to the category of physical being that situates it outside the distinction between living and non-living entities. Second, to consider, on that basis, the relevance of physical being to biopolitics, which has been obscured in Agamben's treatment of the biopolitical. "In our culture," Agamben writes, "the decisive political conflict, which governs every other conflict, is that between the animality and the humanity of man. That is to say, in its origin Western politics is also biopolitics."[5] My argument in this chapter is that not only the distinction between animality and humanity but also the distinction between physical being and animality is crucial to the framework of biopolitics. Recognizing this helps us situate the biopolitical import of nanoscale materials science without relying upon concepts of "life" that are troubled by new materials and without falling back upon precisely those ontological assumptions that, in *The Open*, Agamben's thinking of the biopolitical is meant to critique.

THE LACUNA OF BIOPOLITICS

Distinguishing between modalities of being-in-the-world proper to the stone (physical being), to the animal (life), and to man (Dasein), Heidegger famously claims that the stone is "worldless" (*weltlos*); the animal is "poor in world" (*weltarm*); man is world-forming (*weltbildend*).[6] When he turns to this schema in *The Open*, Agamben devotes some thirty pages of his text to a treatment of Heidegger's distinction between man and animal, but he affords only one sentence to Heidegger's discussion of the stone: "Since the stone (the nonliving being)—insofar as it lacks any possible access to what surrounds it—gets quickly set aside, Heidegger can begin his inquiry with the middle thesis, immediately taking on the problem of what it means to say 'poverty in world'" (*TO* 51). Quickly set aside by Heidegger, the stone, or non-living being, is set aside even more rapidly by Agamben, leaving unexamined a fundamental distinction in Heidegger's text: the distinction between physical being and life. Because he does not interrogate Heidegger's argument that the stone lacks any possible access to what surrounds it, Agamben does not consider the manner in which his own analysis of the distinction between animal and man at once relies upon and occludes the distinction between living and non-living being. At what would seem like an opportune moment to question the frame of Heidegger's three-part schema—a moment at which to pose questions concerning the precise status of "nonliving being," to ask how the conceptual determination of the "nonliving" sets up Heidegger's concept of "life," to analyze the concept of "access" determining Heidegger's denial of world to non-living entities—Agamben sets aside Heidegger's first distinction (stone and animal) in order to focus on his second (animal and man). In doing so he sets aside the problem of what Heidegger calls "physical being" (*FCM* 192).

Agamben's reticence on this point is symptomatic. Throughout his work on biopolitics, from *Homo Sacer* through *The Open*, he is at pains to analyze categorial fissures internally dividing the concept of "life." Yet he never adequately addresses the distinction between living and non-living being upon which that concept is based. In the opening pages of *Homo Sacer*, Agamben introduces distinctions between *zoē* and *bios*, between bare life and qualified life, between "the simple fact of living common to all living beings" and "the form or way of living proper to an individual or group."[7] He also introduces distinctions between various modalities of *bios*: *bios theōrētikos* (the life of the philosopher), *bios apolaustikos* (the life of pleasure), and *bios politikos* (the political life). These distinctions, among others, open up the conceptual terrain of living being upon which Agamben will carry out his analysis of biopolitics: of the sovereign ban upon and concurrent politicization of bare life from and within the polis, of the irreducible indistinction of *zoē* and *bios* characteristic of modern politics, of the anthropological machine producing the concept of "man" through the opposition man/animal, human/inhuman. But Agamben never takes seriously the problem of non-living being; he never attends to the conceptual distinction between life and non-life upon which distinctions among different concepts and modalities of life are implicitly premised.

What this means is that the conceptual determination of the biological demarcating the boundaries of "biopolitics" in Agamben's thought is implicitly unthought. Agamben is a careful thinker of the exception, of the sovereign *exceptio* through which bare life is simultaneously excluded from and thereby included within the order of politics—included in and through its exclusion from the polis. What Agamben does not seem to see or attend to is that the concept of *zoē* or bare life—the simple fact of living common to all living beings—must itself first be constituted by its categorial separation from that which is not living, from non-living being, from what Heidegger terms "physical being." This is also a categorial separation that founds a concept ("life") upon the exclusion of that which is non-living from the extension of that concept, thereby including this exclusion within its own constitution. The category of simple life, of bare life, of *zoē*, and therefore also of the *bios* or qualified life from which it is distinguished, first *rests* upon the distinction between the living and the non-living, between life and physical being. It is thus crucial—by the lights of Agamben's own method—that any consideration of the relation between life and politics, of "biopolitics," examine this distinction as constitutive of the very concept of "life" upon which biopolitics is predicated. "Every limit concept is always the limit between two concepts," notes Agamben in *Homo Sacer*.[8] Yet his analysis always dwells upon one side of a double limit—within and among distinctions between *zoē* and *bios*, between distinct modalities of life—without ever examining the obverse side of the constitution of "life" as a limit concept: its differentiation from that which is non-living.

For his own part, Heidegger at least recognizes this double determination of "life" as a limit concept, even if he does not treat one side of its distinction as thoroughly as he should. Though it is true that he sets the stone aside more quickly than animal or man, his brief attention to the stone's modality of being as a necessary preliminary to distinguishing between modalities of being proper to the animal and man marks the importance, for him, of establishing a distinction between physical being and living being before making a distinction between life and Dasein. For Heidegger, it is clear that this second distinction is logically conditional upon the first. In order to distinguish Dasein from life, as the different modalities of being-in-the-world proper to man and animal, it is first necessary to know what life is. And this requires a prior distinction between living and non-living being, between life and physical being.

Because Agamben does not attend to this distinction, he cannot situate the concept of "the open" in relation to physical being. But in order to gauge the consequences of this lacuna for biopolitical theory, we need to attend more closely to the distribution of Heidegger's categories in his 1929–30 seminar.

STONE AND ANIMAL

Heidegger's distinction between living and non-living being is phenomenological: he thinks of modalities of being-in-the-world in terms of modalities of *access* to world. The stone (the object, the non-living being) is "without access" to world (*FCM* 196), whereas the animal (the living being) is "not without access" (*FCM* 198). According to Heidegger's phenomenological method, it is not physical law that determines the modality of the stone's being. Rather, it is the stone's not-having-access that "makes possible" the sort of being proper to the order of "material nature" that is governed by physical law (*FCM* 191). Likewise, "the living character of a living being . . . the kind of being that pertains to animals and plants" (*FCM* 191) "cannot be explained or grasped at all in physico-chemical terms" (*FCM* 188). Heidegger's initial distinction between living and non-living being will thus operate according to modalities of access rather than through reference to scientific taxonomies.

What is at issue for Heidegger is "the *living character of a living being*, as distinct from the non-living being which does not even have the possibility of dying." The stone is determined as such a being, which "cannot be dead because it is never alive" (*FCM* 179). As a non-living being, the stone is immobile and senseless. It crops up here or there, but it neither accesses nor interacts with the place in which it is:

> It lies upon the earth but it does not touch it. The stone lies on the path. If we throw it into the meadow then it will lie wherever it falls. We can cast it into a ditch filled with water. It sinks and ends up on the bottom. In each case according to circumstance the stone crops up here or there, amongst and amidst a host of other things, but always in such a way that everything present around it remains essentially *inaccessible* to the stone itself. (*FCM* 197)

"World," according to Heidegger, denotes "those beings which are in each case accessible and may be dealt with, accessible in such a way that dealing with such beings is possible or necessary for the kind of being pertaining to a particular being" (*FCM* 196). The stone is worldless insofar as it does not have such access. It is "essentially *without access* to those beings amongst which it is in its own way" (*FCM* 197). Worldlessness "is constitutive of the stone in the sense that the stone *cannot even be deprived* of something like world." And because this not-having-access is constitutive of its modality of being, it would be "beside the point to regard the fact that the stone has no access as some kind of lack." For Heidegger, it is precisely the fact of the stone's not-having-access that "makes possible its specific kind of being, i.e., the realm of being of physical and material nature and the laws governing it" (*FCM* 197). This is the realm of *physical being.*

Unlike the stone, the animal is "capable of behavior" in an environment, but "the behavior of the animal is not a *doing and acting,* as in human comportment, but a *driven performing*" (*FCM* 237). The animal is "poor in world" insofar as it is "*related to other things*—although *these other things are not manifest as beings*" (*FCM* 254). The animal is open to its environment, but this openness is essentially a "captivation" whereby "the animal's *relationality* to other things consists in a kind of *being taken*" (*FCM* 249). Following the research of German biologist Jakob von Uexküll,[9] Heidegger describes the animal environment as a "ring of disinhibitors," of sensory triggers by which the living being is stimulated "without being able to reflect upon" those stimuli (*FCM* 247). The animal is thus "open in captivation" in the sense that it is "taken and captivated by things" while the apprehension of those things is "taken away from the animal" (*FCM* 247). Captivation, argues Heidegger, "is the condition of possibility for the fact that, in accordance with its essence, the animal *behaves within an environment but never within a world.*" The condition of poverty in world resides in the fact that "beings are not *manifest* to the behavior of the animal in its captivation, they are not disclosed to it and for that very reason are *not closed off* from it either. Captivation stands outside this possibility" (*FCM* 239). Just as the stone's modality of being is made possible, negatively, by its not-having-access, that of the animal is enabled by its being-outside-of the manifestness of beings: the living being is "suspended, as it were, between itself and its environment, even though neither the one nor the other is experienced *as* being" (*FCM* 239).

Man, then, is that being, Dasein, for which beings are manifest *as* beings, and this manifestness of beings as beings "belongs to world." For the animal, beings are merely present at hand: "there is no indication that the animal somehow does or ever could comport itself toward beings as such" (*FCM* 253). Dasein is that modality of being through which the scission of the ontological difference occurs. "In the first instance," writes Heidegger,

> I am directed toward beings themselves; I confine myself to their qualities. In the second instance, on the other hand, when I consider beings insofar as they *are* beings, I am not investigating their properties, but take them insofar as they are, with respect to the fact that they are determined by their *being.* (*FCM* 322)

The "as" structure—the exposition of beings *as* beings—is constitutive of world formation. For Heidegger it is thrown projection—the irruptive opening of temporalization proper to being-there—that primordially opens the ontological difference, the letting-prevail of world as the difference between beings and being (*FCM* 362).

Heidegger summarizes this double distinction—between non-living and living being, between living being and Dasein—through two examples. If we consider a worm that flees from a mole and agree that the worm cannot "comport itself toward" the mole in the manner characterizing human being, we might also agree that it nevertheless "*behaves with respect to*" the mole. But a stone, argues Heidegger, "cannot behave in this way" (*FCM* 237). Similarly, if we consider the case of a lizard basking on a rock in the sun, we might agree that "the rock on which the lizard lies is not given for the lizard *as* rock, in such a way that it could inquire into its mineralogical constitution for example." Nevertheless, Heidegger argues, the living being of the lizard differs from the physical being of an object insofar as the lizard "has its *own relation* to the rock, to the sun, and to a host of other things" (*FCM* 197). Note that, for Heidegger, the modality of being proper to the lizard as a living being does not depend upon cognitive, psychological, or affective faculties but rather upon its specifically phenomenological relation to world, stripped of these ontic predicates:

> The lizard basking in the sun on its warm stone does not merely crop up in the world. It has sought out this stone and is accustomed to doing so. If we now remove the lizard from the stone, it does not simply lie wherever we have put it but starts looking for its stone again, irrespective of whether or not it actually finds it. The lizard basks in the sun. At least this is how we describe what it is doing, although it is doubtful whether it really comports itself in the same way as we do when we lie out in the sun, i.e., whether the sun is accessible to it *as* sun, whether the lizard is capable of experiencing the rock *as* rock. Yet the lizard's relation to the sun and to warmth is different from that of the warm stone simply lying present at hand in the sun. Even if we avoid every misleading and premature psychological interpretation of the specific manner of being pertaining to the lizard and prevent ourselves from "empathically" projecting our own feelings onto this animal, we can still perceive a distinction between the *specific manner of being* pertaining to the lizard and to *animals* and the *specific manner of being* pertaining to a material thing. (*FCM* 197–98)

While the stone is merely present at hand, regardless of what is done to it, Heidegger insists that "it is not true to say that the lizard merely crops up as present at hand *beside* the rock, *amongst* other things such as the sun for example, in the same way as the stone lying nearby is simply present at hand amongst other things" (*FCM* 198). For Heidegger, it is this simple presence at hand, without access to sensory stimuli and without the possibility of behavioral response, that is constitutive of the physical being of non-living entities.

In his treatment of Heidegger, Agamben emphasizes Heidegger's inversion of Rilke's treatment of the distinction between man and animal in his Eighth *Duino Elegy*.[10] In Rilke's text it is the animal (*die Kreatur*) that "looks out / into the Open,"[11] while man's eyes are "turned toward the world objects, never outward."[12] In Heidegger's account the animal or living being is merely "open in captivation" (*FCM* 247) to specific sensory disinhibitors, while only Dasein accesses the manifestness or unconcealedness of beings in the Open of the ontological difference, or the letting-be of world. Addressing Heidegger's reversal of Rilke, Agamben's intervention is to situate the intersection of animal and man in Heidegger's analysis of boredom as a fundamental mood or attunement (*Stimmung*). According to Agamben, boredom is a *suspension* of captivation itself (rather than a behavioral *response* to sensory captivation). Boredom is the place at which "human openness in a world and animal openness toward its disinhibitor seem for a moment to meet" (*TO* 62), and "Dasein is simply an animal that has learned to become bored; it has awakened *from* its own captivation *to* its own captivation" (*TO* 70). The point of Agamben's intervention is to deconstruct the constitution of the human through its distinction from the animal and to insist that access to "the open" is not what is proper to man but rather that which transpires within a zone of indistinction proper to "bare life." The conclusion of Agamben's treatment of Heidegger's 1929–30 seminar is that:

> Being, world, and the open are not, however, something other with respect to animal environment and life: they are nothing but the interruption and capture of the living being's relationship with its disinhibitor. The open is nothing but a grasping of the animal not-open. Man suspends his animality and, in this way, opens a "free and empty" zone in which life is captured and a-bandoned [*ab-bandonata*] in a zone of exception. (*TO* 79)

For Agamben, the open is thus a biopolitical problem. The simultaneous division and articulation of the animal and the human by the anthropological machine doubles the simultaneous division and articulation of *zoē* and *bios* through the sovereign ban, and the open is that "central emptiness" (*TO* 92) or "zone of irreducible indistinction"[13] within which the distinction between man and animal, *zoē* and *bios*, paradoxically rests.

Just as Heidegger offers the figure of the lizard to exemplify the distinction between animal and man, Agamben offers an exemplary creature to figure of the zone of indistinction into which bare life falls: he turns to the example of the tick, which had been deployed by von Uexküll in a text published in 1934, five years after Heidegger's 1929–30 seminar.[14] The lifeworld or *Umwelt* of the tick, as described by von Uexküll, consists of a very minimal array of sensory data, or "carriers of significance," by which its tightly constrained behaviors are triggered: the sensitivity of its skin to light, the odor of butyric acid emitted by mammals, the temperature of thirty-seven degrees Celsius corresponding to that of mammalian blood, and the tactile properties of its host's skin. As the tick hangs inert upon the tip of a branch, von Uexküll notes, no stimulus perturbs it. But as a mammal approaches and passes by, the odor, temperature, and haptic properties of the animal "glow like signal lights in the darkness." Brushed off the branch by its passing host, the tick finds its way into the flesh of the mammal's body, drinks its blood, falls to the ground, lays its eggs, and dies. In its sensory and behavioral captivation by a minimal array of environmental triggers, writes Agamben, "the example of the tick clearly shows the general structure of the environment proper to all animals" (*TO* 46).

But Agamben also notes an instance in which the tick becomes an *improper* example of the animal's modality of being. Von Uexküll refers to research conducted in a laboratory in Rostock, where a tick was kept alive for eighteen years without food and "in a condition of absolute isolation from its environment" (*TO* 46)—without any relation, that is, to the only sensory stimuli it is capable of receiving. "What becomes of the tick and its world in this state of suspension that lasts eighteen years?" Agamben asks. "How is it possible for a living being that consists entirely in its relationship with the environment to survive in absolute deprivation of that environment? And what sense does it make to speak of 'waiting' without time and without world" (*TO* 47)? "Without world," Agamben writes. Here he touches upon the threshold between living and non-living being in Heidegger's account, upon the limit between the animal's poverty in world and the stone's worldlessness—a liminal condition that is occupied by the tick in the Rostock laboratory, deprived of any access to an environment. But Agamben fails to explicitly register or analyze the animal's proximity to the stone in this example, and when he later grapples with the significance of von Uexküll's theory, what concerns him is that "the animal can effectively suspend its immediate relationship with its environment, without, however, either ceasing to be an animal or becoming human." Again, he does not register or address its proximity to "becoming stone." For Agamben, the tick in the Rostock laboratory "guards a mystery of the 'simply living being,' which neither Uexküll nor Heidegger was prepared to confront" (*TO* 70). On this point, he is correct. But Agamben is equally unprepared to confront this problem, which is not that of the relation between man and animal but rather the problem of a liminal condition between "life" and "physical being."

Suspended without access to an environment, the worldless existence of the tick in the Rostock laboratory occupies a zone of indistinction between living and non-living being. Indeed, its existence is that of what Agamben calls, elsewhere, "neither an animal life nor a human life, but only a life that is separated and excluded from itself" (*TO* 38). But in what sense is this "life" at all? If the stone, according to Heidegger, "cannot be dead because it is never alive," then the tick, paradoxically, would seem to have become a stone that nevertheless *remains* alive. It attains the state of being without access to an environment, yet it retains the capacity to die. The tick thus exists as bare life, but it does so in the worldless modality of physical being. It is *this* threshold, between life and physical being, that constitutes "the mystery of the simply living being" with which Agamben's later work is so persistently concerned but which he never directly addresses.

We can now approach this threshold from the other side of its limit, at which it is not the living being that exists at the border of non-living being but rather the physical being of the object that exists at the border of "life," or that which is "not-without-access" to world.

NANOTECHNOLOGY AND PHYSICAL BEING

As long as reflection upon the physical being of non-living objects is restricted to the figure of "the stone," as in the case of Heidegger's text, it might seem relatively uncontroversial to claim that the physical being of such an object is determined by its not-having-access to that which surrounds it.[15] But as soon as we engage with objects and materials that would normally be deemed "non-living" yet exhibit precisely those traits of access to environmental stimuli and behavioral response which Heidegger considers proper to "life," the determinations and distinctions of Heidegger's influential seminar need to be reconsidered. Such a reconsideration should ultimately lead us to question Heidegger's determination of non-living being, or physical being, as it applies to *any* material object, including "the stone." This is the case because any categorial determination of physical being—as a modality of being proper to "the object" and strictly distinct from "life"—must apply to any and all non-living beings if it is to warrant such a categorial distinction. If it is possible not just to enumerate exceptions to the determination of such a category but to show how these undermine the very terms upon which the category is thought, then the utility and categorial integrity of the determination itself are in question. The point of considering such exceptions is not merely to challenge Heidegger's schema but rather to explore the consequences of thinking the category of physical being otherwise, for both philosophy and poetics. That is the burden of what follows in this chapter and the next.

Among the primary goals of nanoscale materials science and engineering is the fabrication of physical materials that (to speak in terms of Heidegger's determination of living being) are not-without-access to an environment—that is, materials that relate to their environment in at least the minimal register of von Uexküll's tick: triggered by particular environmental stimuli and behaviorally responsive to those triggers. These are materials that challenge Heidegger's distinction between modalities of non-living and living being. Indeed, the distinction between living and non-living entities is challenged in any number of ways by the capacity of nanotechnologies to image, manipulate, and guide the self-organization of matter on molecular and submolecular scales. In the field of nanobiology, for example, researchers attempt to "assemble artificial cells from scratch using nonliving organic and inorganic materials."[16] But what is at issue apropos of Heidegger's argument is not only the challenge such research might pose to definitions of life predicated upon the *physical constitution* of living bodies, or their "natural" or "artificial" production, but also the challenge posed to determinations of living and non-living being grounded in distinctions between certain capacities, functions, or behaviors correlated to distinct modalities of being-in-the-world. The procedure of nanoscale science and technology is often to study structures and processes of "life"—viral architecture, bacterial self-replication, molecular self-assembly, biological stimulus/response systems—in order to replicate those structures and processes in material contexts that are not confined by the cellular organization or the chemical requisites of the organism.[17] This project relies upon precisely the decoupling of a "mode of being" from any essential determination by physico-chemical structure—precisely the decoupling at stake in Heidegger's phenomenological analysis. Researchers in the field of nanoscience and technology study the physio-chemistry of life in order to replicate the phenomenon of living being by other means.

For example, Minoru Taya, director of the Center for Intelligent Materials and Systems at the University of Washington, studies principles of light and touch sensing in plants so as to apply their trigger and actuation mechanisms to the construction of inorganic materials with the capacity to sense and

Figure 9 Nano skin. Fabricated by Pulickel Ajayan's research team at Rensselaer Polytechnic Institute. Courtesy of Pulickel Ajayan.

respond to their environment. "Biological systems are ideal adaptive structures with smart sensing capabilities," he explains, and understanding these is crucial to "designing adaptive structures and intelligent materials."[18] While Taya's research involves micro- and macroscale materials, such as artificial polymer-based actuators as alternatives to biological, filament-based muscles, Pulickel Ajayan's research team at Rensselaer Polytechnic Institute constructs materials capable of integrating nanoscale fabrication into the sort of biomimetic sensors Taya describes. The "nano skins" fabricated by Ajayan's team are flexible hybrid composite materials consisting of a polymer substrate embedded with organized arrays of carbon nanotubes—cylindrical tubes of hexagonally organized carbon atoms more than ten thousand times smaller than the diameter of a human hair (Fig. 9).[19]

Since these materials function as field emitters, like those used in electronic screens, they are being developed as a template for flexible electronic devices. And because arrays of carbon nanotubes maintain their high conductivity and electrical sensitivity when embedded in polymers, nano skins could be used as adhesive structures, pressure sensors, or gas detectors.[20] Ajayan points out that these materials also constitute a step toward the development of biomimetic sensors, since carbon nanotubes have the capacity to convert mechanical signals into electrical signals—one of the key functions of epidermal cells in plant and animal stimulus/response systems.[21] As Ajayan has demonstrated with a research team at the University of Akron, nanotube arrays embedded in polymers can be used to mimic the action of microscopic sensory hairs on epidermal surfaces, which can carry mechanical stimuli to a receptor cell that then converts those stimuli into an electrical signal, propagating ion flow through neighboring cells and activating a plant's motile action (as in the actuation system of the Venus flytrap, for example).[22] This biomimetic capacity of Ajayan's hybrid polymers is particularly notable when considered alongside the work of MIT engineers on macroscale solid compounds that expand and contract through ion flow. Starting with compounds commonly found in lithium-ion rechargeable batteries, an MIT research team led by Yet-Ming Chiang and Steven Hall has developed prototypes of electrochemically actuated "morphing materials" that offer "a synthetic counterpart to the nastic actuation mechanism in plants."[23]

The capacity of nanoscale materials to operate as mechanical sensors and electrochemical signals—and the potential of coupling of such sensors to actuation mechanisms—is also demonstrated by the DNA-wrapped carbon nanotubes engineered by Michael Strano's research team at the University of Illinois at Urbana-Champaign. The sensors designed by Strano's team consist of a strand of DNA wrapped around a single-walled carbon nanotube "in much the same fashion as a telephone cord wraps around a pencil."[24] Exposed to the ions of certain atoms, negative charges along the strand of DNA are neutralized, altering its shape and reducing its surface area. This shift perturbs the electronic structure of the carbon nanotube, altering its emission energy—a process that can be reversed by removing the ions from the DNA.

Strano's team reports that "the nanotube surface acts as the sensor by detecting the shape change of the DNA as it responds to the presence of target ions," and this response can be measured, enabling the detection of low concentrations of mercury ions in mammalian cells and tissues.[25] Researchers like Deborah Estrin, founding director of the Center for Embedded Networked Sensing at UCLA, seek to integrate such nanoscale devices into "massively distributed collections of smart sensors and actuators embedded in the physical world." Such networks would operate both inside and outside of living bodies, and their continued operation would rely upon "self-configuring systems that adapt to unpredictable environments where pre-configuration and manual intervention are precluded."[26]

The "adaptive structures and intelligent materials"[27] designed by nanoscale materials engineering figure as extremely minimal examples of a capacity for "biomimetic" environmental "access" in physical systems. However, considered at the scale of a project like Embedded Network Sensing, nanoscale engineering also portends a thorough transformation of whatever we consider an "environment" into that which is "not without access" to itself. That is, the prospect of a pervasive integration of nanoscale sensor/actuator systems tends to erase the distinction between the "modality of being" of a living system within an environment and the environment itself. Again: it is not my intention to argue that such devices, or the distributed networks into which they may eventually be embedded in the physical world, are "alive," or that the category of "life" should be expanded to include them. Nor do I seek to demonstrate, through any theory of "life" elaborated on my own terms, that such entities share "the kind of being that pertains to animals and plants." Rather, my argument is that *if* the "way of being" called "life" is to be determined as "*not without access* to what is around it and about it" (*FCM* 198), and *if* that way of being is to be delimited by the material object's "not having access" to its surroundings, then this determination and this delimitation are called into question by the capacities of materials and devices being developed by nanoscale engineering today. The import of this fact is not that nanotechnology portends an epochal rupture in the history of being but rather that the capacities of the objects, systems, and networks it produces encourage us to rethink the categorial determinations through which we have distinguished living being from physical being.

My approach to this problem differentiates my understanding of the conceptual stakes of nanotechnology from Colin Milburn's discussion, in *Nanovision: Engineering the Future*, of what he calls "the postbiological body." Attending to the speculative discourse of the nanotechnology industry, Milburn examines the rhetorical construction of a postbiological future in which the functions and capacities of living beings are artificially produced and distributed among organic, inorganic, and hybrid bodies. This speculative discourse, Milburn points out, "sees the postbiological not as the end of life but rather the condition of 'surplus life,' the 'excessively biological,' or the overflow of life beyond itself."[28] The discursive retention of the categories of life and of the biological to designate that which is beyond or outside the "natural" parameters of those categories serves to reincorporate properties and capacities like auto-motility, self-organization, or adaptive response back *into* the category of life even as these properties or capacities are predicated on entities or materials that might be taken to trouble the constitution of that category. We might call this a rhetorical program of *biological* or *vitalist recuperation*.

The terms "postbiological" and "postvital" themselves indicate an unwillingness to articulate a concept that would specify this "other" modality of being as anything other than "the afterlife of life," as Milburn puts it.[29] The subtitle of Richard Doyle's book, *Wetwares: Experiments in Postvital Living*, makes this clear. And while Milburn's analysis of the speculative rhetoric surrounding nanotechnology is well aware of its internal contradictions, he nonetheless leaves the vitalist or biocentric conditioning of that rhetoric intact. For Milburn, the postbiological "would describe the profligate production of alternative modes of living—machinic life, technogenic life, nonorganic life, nanotech life, quantum life, galactic life, and so forth—as a direct consequence of the singularity of 'life itself' giving way to its outside," and this would amount to "an issuing of 'lives' in the place of 'life.'"[30] What is impeded by this account is the possibility of thinking the "outside" of life as anything *other* than life. My argument is that in order to overcome this rhetorical and conceptual difficulty, what is necessary is not an expansion of the category of life but, rather, a rethinking of physical being. This requires us to challenge not only the logic of the "postbiological" but also the biopolitical.

At the intersection of the materials and devices engineered by nanotechnology—which are not-without-world—and the becoming-stone of the animal exemplified by the worldless isolation of the tick in the Rostock laboratory, Heidegger's triple thesis undergoes an implosion on the outside of the binary distinction between man and animal that Agamben deconstructs in *The Open*, and therefore on the outside of "life." As Agamben might point out, were he attentive to this problem, this outside is also the inside of that distinction, insofar as it is the ground upon which the *zoē* common to man and animal, to living beings in general, can be delimited in the first place. It is thus Agamben's determination of the category of "biopolitics" itself that needs to be rethought, insofar as it what it leaves unexamined is the delimitation of the biological upon which that category is predicated.

THE INORGANIC OPEN

My argument has been that Agamben's framing of the biopolitical participates in the operations of "the anthropological machine" that he purports to critique in *The Open*. This is to argue that what Agamben calls the anthropological machine not only functions by separating man from animal by producing a zone of indistinction in which we find "neither an animal life nor a human life, but only a life that is separated and excluded from itself—only a bare life" (*TO* 38). The anthropological machine *also* separates living being from physical being by producing a zone of indistinction between these categories, into which their separation falls. Thus, that which the anthropological machine produces as "bare" or *haplōs*—the isolation of that which is pure or simple or unqualified, a minimal term cast outside of "world"—would not be a generic modality of living being (bare life) but rather the "worldless" isolation of *physical* being as that which must be separated and expelled, in the first instance, in order to define the domain of "life" within which man can be separated from animal.[31] It is remarkable that this point has gone unrecognized throughout the articulation of biopolitical theory from Foucault to Esposito.[32]

As it concerns the relation of biopolitics to technology, we can trace the effects of this oversight in Foucault's foundational articulation of the problem of biopolitics in the final lecture of *Society Must Be Defended*:

> This excess of biopower appears when it becomes technologically and politically possible for man not only to manage life but to make it proliferate, to create living matter, to build the monster, and, ultimately, to build viruses that cannot be controlled and that are universally destructive. This formidable extension of biopower, unlike what I was just saying about atomic power, will put it beyond all human sovereignty.[33]

What is missed here is that the technical capacity to "create living matter"—"not only to manage life but to make it proliferate"—is precisely that which calls into question the very delimitation of the category of "life." To create living matter—artificial cells, for example—is at one and the same time to undo the category of living matter by denaturalizing and destroying its criteria. That is: the possibility of the *artificial* production of living being is paradoxically due to its indistinction from physical being, the "zone of indistinction" through which the separation of living and non-living being is conceptually produced in the first place. If Heidegger's *phenomenological* determination of the distinction between life and physical being, as discrepant modalities of being-in-the-world, allows him to avoid definitions premised upon the ontic, physical constitution of entities, we have seen that his phenomenological distinction is also insufficient to uphold the categorial separation of living and non-living being.

At its apogee in *The Open*, the failure of biopolitical theory to encounter the problem of physical being remains legible even in Timothy Campbell's recent study, *Improper Life: Technology and Biopolitics from Heidegger to Agamben*. Despite Campbell's attention to "improper life," his study does not take up the pertinence of technics to biopolitics in relation to the distinction between living and non-living being. Rather, Campbell's reflections are concerned with the "thanotopolitical" import of technology: its pertinence to the relation between life and death. The problem is that a consideration of the biopolitical import of technics oriented by the relation between life and death cannot touch upon the very *constitution* of the categories of "life" and "death," insofar as both of these depend upon the prior distinction between living being and physical being (as Heidegger notes, a stone "cannot be dead because it is never alive" [*FCM* 179]). Thus, in his chapter titled "Heidegger, Technology, and the Biopolitical,"[34] Campbell does not touch upon the aporia at the heart of biopolitics per se. The distinction between living and non-living being remains outside the purview of his investigation, although it determines and delimits that which is *within* the purview of his investigation.

It is thus on the *outside* of biopolitical theory, via the work of Bernard Stiegler and Jean-Luc Nancy, that we might begin to think through this problem. Both of these thinkers, in different registers, return us to the problem of "the stone" in Heidegger's work.

In volume one of *Technics and Time*, Stiegler interrogates "the invention of the human" via theories of technological evolution, analyzing "the pursuit of the evolution of the living by other means than life."[35] At the core of this investigation is the constitution of the categories of man and animal through the relation between life and the physical object, living being and physical being. Stiegler shows that what resides in the zone of indistinction between man and animal is the stone, in its function as technical object. It is with *flint* that memory, language, thought come to be carved into a mineral substrate, and thereby emerge in the first place through a process of exteriorization. And it is through this process of exteriorization, as technical recording, or arche-writing, that we encounter the paradox of "the technical inventing the human" and "the human inventing the technical" (*TT* 137). The distinction between man and animal derives from this "default of origin" whereby cortical evolution is codetermined by a process of technical evolution through which the relational differentiation of animal and man is mediated by physical being: "Corticalization," Stiegler writes, "is effected in stone" (*TT* 134).

Stiegler's engagement with Heidegger in *Technics and Time* is primarily devoted to the existential analytic of *Being and Time*, but the consequences of his analysis for the distinctions developed in the 1929–30 seminar are clear. As "the fundamental structure of world-formation" (*FCM* 362), it is *projection* that primordially opens access to being *as* being. And it is this opening of the ontological difference that differentiates the Dasein of man from the life of the animal. Projection involves a "leaping ahead of oneself" constitutive of *anticipation* and a "falling back into everydayness" conditioned by Dasein's *thrownness* into a context of factical existence.[36] Stiegler shows that both anticipation and the constitution of facticity depend upon a technical exteriorization of memory that founds historicity and opens futurity. Thus, technics is a condition of possibility for projection, such that Dasein's world-forming essence—along with the distinction of Dasein from life—emerges from the structural coupling of life with "nonorganic organizations matter" (*TT* 17). It is not that tools are invented through man's world-forming capacities, which distinguish man from animal, but rather that "the first flint knapped tool" (*TT* 135) *coevolves* with life in a process of world-formation that is a condition of possibility for Dasein and for the distinction of man from animal. For Stiegler, world-formation is an irreducibly double phenomenon, a "double plasticity" (*TT* 142) by which both the inorganic object and life form one another through a "meeting of matter" (*TT* 141). In *Technics and Time*, the essence of man falls into the fracture *between* animal and man, the default of origin within which we find a coevolutionary relation between living and non-living being: "a double emergence of cortex and flint" (*TT* 155). The deconstructive question posed by Stiegler's analysis for the ontological schema of Heidegger's 1929–30 seminar is: "What plasticity of gray matter corresponds to the flake of mineral matter" (*TT* 135)?

For our purposes here, however, there is a problem with Stiegler's account of the invention of the human. Stiegler's analysis of technical coevolution proceeds on the basis of a distinction between the "inorganic organized being" (the technical object) and the "inorganic beings of the physical sciences" (*TT* 17). Beginning his study with a critique of the Lamarckian distribution of physical bodies into two classes—"the non-living, inanimate, inert" and the organic being (*TT* 1)—Stiegler elaborates the ontology of technical objects as "a third genre of 'being'" (*TT* 17). Technical objects, Stiegler argues, "have their own dynamic when compared with that of either physical or biological beings, a dynamic, moreover, that cannot be reduced to the 'aggregate' or 'product' of these beings" (*TT* 17). "There is a *historicity* to the technical object," he writes, "that makes its descriptions as a mere hunk of inert matter impossible. This inorganic matter organizes *itself*" (*TT* 71). Thus Stiegler reinscribes a distinction between inorganic matter which has a historicity (technical object) and inorganic matter which is merely "inert." The category of the technical object seems to deconstruct the Lamarckian distinction between "either physical or biological beings" by introducing a third term, but it does not *displace* this distinction by unsettling the initial determination of physical beings upon which it depends; it merely distinguishes and separates technical objects *from* those beings. For Stiegler, the technical object is world-forming, but the merely inert physical object remains worldless. It suffices to note, however, that this distinction reproduces the default of origin between animal and man at another level:

that of the processual differentiation of the technical object from merely inert inorganic matter.

It is in order to address the problem of the "mere hunk of inert matter" that Jean-Luc Nancy returns to the schema of Heidegger's 1929–30 seminar in *The Sense of the World*. In a chapter called "Touching," Nancy challenges Heidegger's determination of physical being directly, quoting his text at length:

> The stone is without world. The stone is lying on the path, for example. We can say that the stone is exerting a certain pressure upon the surface of the earth. It is "touching" the earth. But what we call "touching" here is not a form of touching at all in the stronger sense of the word. It is not at all like *that* relationship which the lizard has to the stone on which it lies basking in the sun. And the touching implied in both cases is above all not the same as *that* touch which we experience when we rest our hand upon the head of another human being. . . . Because in its being a stone it has no possible access to anything else around it, anything that it might attain or possess as such.[37]

Nancy's questions for Heidegger are as follows:

> Why, then, is "access" determined here *a priori* as the identification and appropriation of the "other thing"? When I touch another thing, another skin or hide, and when it is a question of this contact or touch and not of an instrumental use, is it a matter of identification and appropriation? At least, is it a matter of this first of all and only? Or again: why does one have to determine "access to" *a priori* as the only way of making-up-a-world and of being-toward-the-world? Why could the world not also *a priori* consist in being-among, being-between, and being-against? In remoteness and contact without "access"? Or on the threshold of access? (*SW* 59–60)

These questions touch upon many of Nancy's primary philosophical concerns, particularly as these have developed through his long engagement with Heidegger: above all, his thinking of "being singular plural," which grants conceptual priority to the concept of being-with in Heidegger's existential analytic, as well as his critique of Heidegger's proper/improper distinction and its attendant thematics of appropriation.[38] Heidegger's thinking of access in terms of *possession* and of modalities of being-in-the-world in terms of *having* access to world results in an elision of what Nancy terms the "threshold of access": the limit or surface at or along which touch transpires.

The singular plurality of being, for Nancy, requires a thinking of world oriented by the fragmentary *effraction* of being-with ("being-among," "being-between," "being-against") such that it is the liminal contact of bodies, *partes extra partes*, that makes up what Nancy calls "the sense of the world." In failing to situate the locus of sense at "the *contact* of the stone with the other surface, and through it with the world as the network of all surfaces," Nancy argues that Heidegger "misses the surface in general" (*SW* 61). Nancy's ontological dislocation of subject and object through a thinking of *bodies* situates *ekstasis* not only as the temporalizing orientation of Dasein's thrown projection, qua being-toward-death, but as the very "extremity" of bodies, the threshold of their liminal contact along which existence takes place in the modality of spacing, *espacement*.[39] Nancy thinks "projection" as the spacing of physical bodies along and through the constitution of their limits, each in relation to the others. "In itself," Nancy writes, "the thing is 'toward' the *other things*," such that "the *différance* of the toward-itself, in accordance with which sense opens, is inscribed *along the edge* of the 'in-itself'" (*SW* 62–63). Threshold, surface, limit, edge: the articulation of the in-itself along the extremity of bodies is thought as the *opening* of world.

We can say, then, that Nancy's critique of Heidegger's treatment of the stone results in a thinking of the *inorganic open*. "All bodies," he writes, "each outside the others, make up the inorganic body of sense," and "sense" is "matter forming itself, form making itself firm" (*SW* 63). The "outside" of "each outside the others," in Nancy's thought—which is also the outside of "life"—replaces the predicative having of world, in Heidegger, such that for Nancy, "world is nothing other than the touch of all things."[40]

The open, from such a perspective, is not circumscribed as the "central emptiness" constitutive of the division of life into man and animal, *bios* and *zoē*. Rather, the open is thought as spacing, articulation, and delimitation taking place between bodies. The open is thought as the modality of *physical being*, no longer determined as a privative modality of worldlessness characterizing that which is inert and without-access. Rather, physical being is that modality of being through which the opening of world takes place *outside* the categorial specification of bodies as stone, animal, and man. That is, physical being is not that modality of being-in-the-world restricted to "the stone" insofar as it is an inert object. Rather, physical being is that modality of being which traverses distinctions between stone, animal, and man. This understanding of physical being offers an alternative to the biocentrism of Agamben's thinking of the open and to the implicit anthropocentrism of Heidegger's.

"The 'open,'" writes Nancy, "is neither the vague quality of an indeterminate yawning nor that of a halo of sentimental generosity. Tightly woven and narrowly articulated, it constitutes the structure of sense qua sense of the world" (*SW* 3). Thus the open is a category that bears upon the constitution of *form* and the articulation of *structure.* But we have yet to understand how the inorganic open, the thinking of the open according to the modality of physical being, gives way onto a thinking of form, of articulation, of structure—of "sense" qua fabrication. Perhaps such an understanding of the open articulates the grounds upon which technoscience and poetry might be thought and situated at the limits of fabrication. To grapple with this possibility, we can now turn to Charles Olson's thinking of "the open" and its relation to his poetics of "objectism," also known as "open field poetry."

2

OBJECTISM: CHARLES OLSON'S POETICS OF PHYSICAL BEING

It is a matter, finally, of OBJECTS.

—Charles Olson, "Projective Verse"

The final facts are, all alike, actual entities; and these actual entities are drops of experience, complex and interdependent.

—Alfred North Whitehead, *Process and Reality*

OLSON'S OBJECTISM, WHITEHEAD'S COMMON WORLD

Charles Olson's poetry is usually classified under the label "Projective Verse," "Open Field Poetry," or "Black Mountain Poetry," and Olson is typically grouped in these categories with other mid-twentieth-century American poets like Robert Creeley, Robert Duncan, and Denise Levertov. But what if Olson's work was remembered today by that other name that he *also* gave to his poetic practice: *objectism*? Would we think of Olson any differently if that term were the first association that came to mind when we heard his name? How might that association change the way we situate his concerns with respect to those of his contemporaries? How might it alter our understanding of the pertinence of his work to recent poetry and poetics? And how might a reconsideration of "objectism" transform our reception of the sense of the world to be gleaned from Olson's poetry?

Before turning to Olson's poetics, let me pause to distinguish my approach to the philosophical problem of the object from that of two other theorists: Graham Harman and Bruno Latour. In this chapter I try to understand Olson's claim that "man is himself an object"[1] while elaborating the non-anthropocentric, anti-vitalist implications of his poetics. But this effort is not aligned with either "object-oriented philosophy" or "actor-network theory." In the work of Graham Harman and his acolytes, objects are understood as constitutively non-relational. The central claim of "OOO" is that objects are discrete insofar as they are "withdrawn" from relation; they are supposedly "devoid of all relation"[2] such that "the actuality of the object belongs always and only to a vacuum."[3] Although Harman attempts to supplement this supposed withdrawal of objects from relation through a theory of what he calls "vicarious causation,"[4] the fact remains that the integrity of discrete objects, throughout his work, depends upon their "vacuum-sealed isolation."[5] The irony of this position is that it implicitly relies upon the epistemological anthropocentrism it is meant to oppose. If what Harman calls "real objects" are supposedly withdrawn from all relation, then the only beings for which they exist at all are those who can articulate the *idea* of their non-relational being and thus dogmatically posit their existence in theory: us. Since they do not enter into causal relations, what Harman calls real objects are strictly impossible to separate from the idea of their existence: they do not exist for any beings other than those who can say they have an idea of them. Thus the avowed anti-materialism of so-called "object-oriented ontology" relies upon an anthropocentric idealism, whose only criterion for the constitution of "real objects" is the dogmatic metaphysical positing of their existence. In short, "object-oriented ontology" is not oriented toward objects at all; it is oriented toward a subjective idea of non-relation, the relational positing of non-relation.[6]

On the other hand, Latour's actor-network theory articulates a relational theory of "actants" composing "hybrid" societies that traverse the distinction between nature and culture. Yet this theory of agency, in its zeal to follow the distribution of "actor-networks" across the binaries of "the modern settlement" by attending to "nonhuman agents," ends up installing a monolithic distinction between "humans" and "nonhumans" at the center of its enterprise.[7] While the category of the "agent" supposedly flattens the field of the social by leveling ontological hierarchies among entities, the agents forming "collectives" are constantly qualified as either "human agents" or "nonhuman agents," thus reinscribing a single categorial split mediated by the concept of agency. Conceptually determined as "nonhuman agents," *objects* are thus determined by an anthropocentric negation that only draws them back into the conceptual primacy of the human as an organizing category.[8] In this sense, the "nonhuman turn" of recent theory[9] is a contradiction in terms; it reasserts "the human" as the central category in relation to which modalities of being are determined. While "object-oriented philosophy" gives us an idea of "real objects" that have no relation to anything but our idea of them, "actor-network theory" gives us a valorization of "nonhuman agency" that, by negation, conceptually subsumes objects within the organizing category of the human. In what follows, on the contrary, I elaborate the influence of Whitehead's *relational* theory of actual entities on Olson's poetics and I develop a materialist account of Olson's "objectism" as a "stance toward reality" that situates the body and the poem in "the larger field of objects," rather than situating the larger field of objects within the parameters of a distinction between the human and the nonhuman.

Olson defines the "stance toward reality" that he calls "objectism" in the second half of his 1950 manifesto, "Projective Verse":

> Objectism is the getting rid of the lyrical interference of the individual as ego, of the "subject" and his soul, that peculiar presumption by which western man has interposed himself between what he is as a creature of nature (with certain instructions to carry out) and those other creations of nature which we may, with no derogation, call objects. For a man is himself an object, whatever he may take to be his advantages.[10]

This is a well-known passage in Olson's writing. But if the term "objectism" nonetheless remains much less familiar than the other terms by which Olson's poetics have been identified, that is largely due to the fact that both Olson's supporters and his detractors have routinely positioned his theory of poetic form as constitutively bound to a theory of the organism, the biological body, and indeed "life," granting these terms a priority in his thought that might seem at odds with the proposition "man is himself an object."

Joseph Conte, for example, aligns Olson with the tradition of organicist poetics stemming from Coleridge, juxtaposing a lineage that values "form as proceeding" against a tradition of procedural or constraint-based methods valuing "form as superinduced."[11] Marjorie Perloff, in an essay to which I will return, takes Olson's breath poetics as a primary exemplar of the 1950s vogue for "the natural look" in poetry, linking his sensibility in this regard with that of Allen Ginsberg, Denise Levertov, Robert Duncan, and the Robert Lowell of *Life Studies*.[12] Steve McCaffery finds in Olson's poetics "a surprising belatedness, a residual organicist romanticism,"[13] while this same supposed "organicist romanticism" is celebrated by Paul Nelson, who begins an essay by replacing the iconic headings of Olson's "Projective Verse" manifesto—"(projectile (percussive (prospective"—with his own: "(organismic (holistic (exploratory."[14] In *The Poetics of the Common Knowledge*, Don Byrd's analysis of the influence of cybernetic theory on Olson's poetics ultimately aligns his writing with "the logic of the living" and the principles of "biological autonomy" propounded by Humberto Maturana and Francisco Varela's theory of autopoietic systems.[15] And in his 1979 volume *Enlarging the Temple*, Charles Altieri provides a formulation that will serve as a synecdoche for the emphasis that both Olson's friends and enemies have placed upon the primacy of the organic, the biological, and the vital in his poetics. Olson's method, Altieri writes, "is imaginative participation in what all beings share by being alive."[16]

Of course, there are many passages in Olson's prose and poetry confirming his investment in the organic, the biological, and the vital. But I will argue that the import of Olson's "objectism" as a "stance toward reality" is that these modalities of existence interest him insofar as they qualify particular objects—not insofar as they establish criteria for what is *not* an object. In his essay *Proprioception*, for example, the proprioceptive sense of our disposition in space and of the body's interiority is characterized by Olson as "the data of depth sensibility/the 'body' of us as object."[17] Though it may be a faculty particular to certain highly evolved organisms, Olson's definition of proprioception as our sense of "the 'body' of us as object" suggests that those operations of the body peculiar to its organic functioning should be considered a *subset* of its facticity as an object. In a word, Olson's understanding of "the 'body' of us as object" is at odds with any ideology or theory for which organic structure or biological functions essentially *distinguish* living bodies from objects. It is at odds, in particular, with the sort of organicism for which the holistic systematicity and continuous self-production of the organism radically distinguish it from entities lacking those qualities—entities that are thus categorically differentiated from organisms and determined as "objects" by *virtue* of that distinction. In a letter to Cid Corman, Olson rejects the principle "that life (what is 'human') is an absolute." Rather, he insists,

> "Life," that dirty capital (doric, corinthian OR iambic—IS RELATIVE to conditions of REALITY (as distinguished from, as *ahead* of life "human life," at any given time:::life, in this sense, is a stop to consolidate gains already being pushed beyond by the reality instant to you or to any man who is *pushing*.[18]

"Life," in other words, does not constitute a starting point or first principle for Olson. He seems to situate any position that grants primacy to Life, "that dirty capital," as a form of classicism: "doric, corinthian OR iambic." Reality, for Olson, is "*ahead* of life" but also ahead of human agency ("any man who is *pushing*"). Life is relative to conditions of reality insofar as these are prior to and encompassing of any living body or human subject. In "Projective Verse," Olson will insist that the reality of either the body or the poem "is a matter, finally, of OBJECTS"—or, more precisely, of "objects in field." In a famous passage, he insists that

> every element in an open poem (the syllable, the line, as well as the image, the sound, the sense) must be taken up as participants in the kinetic of the poem just as solidly as we are accustomed to take what we call the objects of reality; and that these elements are to be seen as creating the tensions of the poem just as totally as do those other objects create what we know as the world.[19]

Thus what has been lost in representations of Olson as a purveyor of "the natural look," as a champion of "organic form," as an apostle of the "lived body," or as an antecedent to "the logic of the living" articulated by autopoietic theory is precisely his rejection of "being alive" as a criterion for participation in "what all beings share." In "Projective Verse," Olson draws our attention not to what all beings share by being alive but rather to the "secrets objects share."[20]

This distinction matters because it bears upon the relation between "life" and "physical being" we took up in the previous chapter and upon the relation of Olson's work to the larger study of "the limits of fabrication" in which we are engaged. When Altieri characterizes Olson's method as "imaginative participation in what all beings share by being alive," he performs a very common gesture whose relevance exceeds his misrepresentation of Olson's poetic project, projecting "being alive" into "all beings" as the basis of our "imaginative participation" in what they share. Indeed, this gesture is the ground of the pathetic fallacy as a rhetorical device. It is not "imaginative participation *in* what all beings share" that is at issue; rather, the constitutive act of "imaginative participation" involved in Altieri's formulation is not so much a matter of "sharing" something *with* all beings as of imputing *to* them our own sense of "being alive." The compensatory attribution of "life" to "all beings" in order to constitute a common world—something all beings share—will be important precisely insofar as one actually understands physical objects in more or less the same privative terms as Heidegger: inert, static, closed, isolated, without world. This is why vitalism has always been merely the flipside of an impoverished understanding of *physical being*: "the role of life," says Bergson, "is to insert some *indetermination* into matter."[21] Or, more recently, in an effort to break with this "habit of parsing the world into dull matter (it, things) and vibrant life (us, beings)," Jane Bennett will speak of "vital materiality" and insist upon "the life of things"—a discourse whose only defense against accusations of anthropomorphism is to affirm the latter as a method.[22] So long as one remains incapable of thinking physical being per se as anything other than worldless closure, and incapable of thinking activity, sense, or the open as anything other than *life*, one will have to rely upon the latter as the ground of any participation in a common world. These are the sort of incapacities that Olson is concerned to avoid, as he makes clear in his letter to Corman.

Representations of physical objects as inert and inactive are first and foremost a matter of *scale*: they are a cognitive by-product of the dominant role played by middle-size objects ("the stone," for example) in human perceptual systems and the codification of relations between such objects by classical mechanics. This is why nanoscale materials research and fabrication is important to our consideration of physical being. In fact, we have known since the elaboration of quantum mechanics that, at sufficiently small scales, material behavior differs markedly from that which we seem to observe in the commonsense objects of normal sense experience. The theory of quantum mechanics, however, seems largely irrelevant to our experience of macroscale objects, insofar as a sofa or table does not exhibit the same behavior as a subatomic particle. But as nanotechnologist and physicist Richard A. L. Jones points out,

> like much received wisdom, there is a kernel of truth in this, surrounded by much that is misleading. The real situation is much more complicated. To start with, some very familiar, everyday properties of the macroworld can only be properly understood in terms of quantum mechanics. Why metals conduct electricity, why magnets attract iron, why leaves are green . . . classical mechanics provides no explanations at all for these questions, which can only be more thoroughly understood in terms of quantum mechanics.[23]

Nanoscale fabrication not only operates on a scale at which quantum mechanics are relevant to the most basic engineering problems; it also attempts to engineer materials in such a way as to incorporate properties peculiar to matter at molecular and submolecular scales (not only quantum mechanical behavior but also molecular self-organization, Brownian motion, surface tension, reduction of grain boundaries) into the structural composition of multiscale systems.[24] "At the nanoscale," write Mark Ratner and Daniel Ratner,

> the most fundamental properties of materials and machines depend on their size in a way they don't at any other scale . . . the nanoscale is unique because it is the size scale at which familiar day-to-day properties of materials like conductivity, hardness, or melting point meet the more exotic properties of the atomic and molecular world such as wave-particle duality and quantum effects.[25]

In a nanoscale wire or circuit, for example, the relation of current, voltage, and resistance formalized by Ohm's law is altered if a wire is only one atom wide, such that electrons have to traverse it one by one.[26] More legibly, perhaps, the development of "smart" windows capable of reflecting sunlight on hot days while letting in heat during colder weather relies upon the nanoscale preparation of chemical compounds that are transparent to infrared light at lower temperatures but undergo a phase transition to begin reflecting infrared when they heat up past a certain point. By preparing the compound (vanadium oxide) as a nanomaterial instead of in bulk, materials scientists have managed to lower its trigger point for this phase transition from 153 degrees Fahrenheit to 90, making this property useful for the treatment of windows exposed to high temperatures (Fig. 10).

"What we found," states the head of the research team, Sarbajit Banerjee, "is an example of how much of a difference finite size can make. You have a material like vanadium oxide, where the phase transition temperature is too high for it to be useful, and you produce it as a nanomaterial and you can then use it right away."[27] If materials science and engineering investigates and fabricates materials based on the relationship between structure and properties, nanoscale materials research and fabrication is concerned with translating what we *know* about the peculiar properties of materials at the nanoscale into transformations of what materials can *do* at the macroscale.

One of the fundamental projects of Olson's poetics was to incorporate his understanding of post-classical science into a "stance toward reality" that would bear upon his compositional practice—his practice of fabrication. In an essay on Melville, "Equal, That Is, to the Real Itself," Olson assesses the consequences of breakthroughs in geometrical and physical theory initiated in the nineteenth century. "All things did come in again, in the 19th century," he writes:

> An idea shook loose, and energy and motion became as important a structure of things as that they are plural, and, by matter, mass. It was even shown that in the infinitely small the older concepts of space ceased to be valid at all. Quantity—the measurable and numerable—was suddenly as shafted in, to any thing, as it was also, as had been obvious, the striking character of the external world, that all things do extend out. Nothing was now inert fact, all things were there for feeling . . . and man, in the midst of it . . . was suddenly possessed or repossessed of a character of being, a thing among things, which I shall call his physicality. It made a re-entry of the universe. Reality was without interruption, and we are still in the business of finding out how all action, and thought, have to be refounded.[28]

Figure 10 Different morphologies of nanostructured vanadium oxide (VO2) attained by varying hydrothermal growth conditions. Scales are in micrometers (um) and nanometers (nm). Courtesy of Sarbajit Banerjee.

"Nothing was now inert fact," Olson writes, since it was "shown that in the infinitely small the older concepts of space ceased to be valid at all." For Olson, post-classical science necessitates a reconsideration of what he terms *physicality* as "a character of being," and this reconsideration was motivated by *scale*—by what we have learned about "the infinitely small."

In the wake of Riemannian geometry, relativity theory, and quantum mechanics, Olson notes that "energy and motion" became "as important a structure of things" as mass, plurality, and spatial extension. In the midst of these discoveries, Olson finds that man "was suddenly possessed or repossessed of a character of being, a thing among things, which I shall call his physicality." *Physicality*, rather than "being alive," is that character of being which establishes the basis of a common world in which qualities of indetermination, motility, and self-differentiation no longer distinguish "life" from "inert matter." In this context, Olson holds that "we are still in the business of finding out how all action, and thought, have to be refounded."[29] In *The Special View of History*, Olson's thesis is that humanity has been "estranged from the familiar" by Socratic rationalism, Kantian idealism, and Newtonian physics. In the wake of this estrangement, he writes, "it has been the immense task of the last century to get man back to what he knows. I repeat the phrase, *to what he knows.* For it turns out to coincide exactly with that other phrase: *to what he does.*"[30] Olson insists that "what you do is precisely defined by what you know," and the question with which he closes his lectures in *The Special View of History* is the following: "After all the 'thought,' what act? . . . what shall we do?"[31]

By the time he delivered *The Special View of History* in 1956, Olson's primary guide to what we know in the wake of post-classical science—to the forms of "thought" this knowledge made available—was Alfred North Whitehead.[32] Whitehead's project in *Process and Reality* was to clarify, by means of a systematic philosophy, the "scheme of ideas" implicit in the scientific breakthroughs of the late nineteenth and early twentieth centuries.[33] The reason that *Process and Reality* resonated so strongly with Olson when he read it in 1955 (already having read Whitehead's *Science and the Modern World* before writing "Projective Verse") is that Whitehead's effort to systematically formulate "what we know" corresponded to Olson's effort to answer the question "what shall we do?" by constructing a poetics, a practice of fabrication, responsive to the conditions of "physicality" described by Riemann, Einstein, Bohr, Heisenberg, and Schrödinger. Both Olson's poetics and Whitehead's speculative philosophy respond to the failure of the doctrine of "simple location" in post-classical science. Against the vulgar materialist principle that "you can adequately state the relation of a particular material body to space-time by saying it is just there, in that place,"[34] Whitehead's post-classical materialism "involves the entire abandonment of the notion that simple location is the way in which things are involved in space-time." Indeed, according to Whitehead, "every location involves an aspect of itself in every other location" (*SMW* 91). The difficulty of grasping the sense of *physical being* at stake in both Olson's objectism and Whitehead's process philosophy is that of holding together this constitutive relationality of "a particular material body" with its determinacy as a coherent unity (an object).

Like Jean-Luc Nancy, Whitehead attempts to think "the open" in the modality of physical being, and his effort to conceptualize a "common world" depends upon thinking both the global relationality and local determinacy of what he calls, alternately, "actual entities" or "actual occasions." "The notion of a 'common world,'" he writes,

> must find its exemplification in the constitution of each actual entity, taken by itself for analysis. For an actual entity cannot be a member of a "common world" except in the sense that the "common world" is a constituent in its own constitution. It follows

> that every item of the universe, including all other actual entities, is a constituent in the constitution of any one actual entity. (*PR* 48)

In his 1920 lectures collected as *The Concept of Nature*, Whitehead will argue that "an object is ingredient throughout its neighborhood, and its neighborhood is indefinite," such that "finally we are driven to admit that each object is in some sense ingredient throughout nature; though its ingression may be quantitatively irrelevant in the expression of our individual experiences."[35] The notion of such degrees of relevance in relations among objects is what rescues actual entities from being "undifferentiated repetitions, each of the other, with mere numerical diversity" (*PR* 148). Whitehead thus offers his "principle of intensive relevance":

> The principle asserts that any item in the universe, however preposterous as an abstract thought, or however remote as an actual entity, has its own gradation of relevance, as prehended, in the constitution of any one actual entity: it might have had more relevance; and it might have had less relevance, including the zero of relevance involved in the negative prehension; but in fact it has just *that* relevance whereby it finds its status in the constitution of *that* actual entity. (*PR* 148)[36]

Thus actual entities or occasions differ among themselves as a result of the gradations of relevance of the determinate relations by which they are constituted, and any particular common world is at once locally *specific* to particular gradations of relevance among particular actual entities and *open* insofar as those entities are, in some sense, "ingredient throughout nature." Particular worlds are topologically specific, we might say, and their local specificity does not preclude their openness with the larger "common world" in which they participate. What is constitutive of participation in the common world is the determinate actuality of beings:

> "Actual entities"—also termed "actual occasions"—are the final real things of which the world is made up. There is no going behind actual entities to find anything more real. They differ among themselves: God is an actual entity, and so is the most trivial puff of existence in far-off empty space. But, though there are gradations of importance, and diversities of function, yet in the principles which actuality exemplifies all are on the same level. The final facts are, all alike, actual entities; and these actual entities are drops of experience, complex and interdependent. (*PR* 18)

What puts beings in common is their complex interdependency, which Whitehead thinks in terms of "experience" because it involves *relations* among entities and occasions, their mutual "prehensions" as constituents in the composition of relational determinacy. Because the process of this composition is ongoing—

because the prehensions of every actual entity or occasion reach beyond its own determinacy even as it is determined by prehensions—there is no totality of beings in Whitehead's cosmology, no ultimate subsumption of a multiplicity under the form of the One. "The community of actual things," Whitehead writes, "is an incompletion in process of production" (*PR* 214–15).

How, then, might an understanding of the relation between Whitehead's theory of the common world and Olson's objectism help us discriminate Olson's essential concerns from the mid-twentieth-century poetics of "organic form" with which he is often associated? And how might doing so help us understand his effort to elaborate a poetics, a practice of fabrication, which situates the determinacy of physical being within what he calls an open field?

OBJECTISM VERSUS ORGANICISM

In her 1991 volume, *Radical Artifice: Writing in the Age of Media*, Marjorie Perloff heads a chapter titled "The Return of the (Numerical) Repressed: From Free Verse to Procedural Play" with an epigraph from Denise Levertov's essay "Some Notes on Organic Form." She then begins her chapter with the following characterization of the mid-twentieth-century *doxa* of the New American Poetry: "Free verse = freedom; open form = open mind, open heart: for almost half a century, these equations have been accepted as axiomatic, the corollary of what has come to be called, with respect to poetic language, the 'natural look.'" To this effect, Perloff immediately cites Charles Olson, Allen Ginsberg, and Robert Lowell:

> "The line comes (I swear it)," announced Charles Olson in "Projective Verse" (1950), "from the breath, from the breathing of the man who writes, at the moment that he writes." "Trouble with conventional form (fixed line count & stanza form)," said Allen Ginsberg in 1961, "is, it's too symmetrical, geometrical, numbered and pre-fixed—unlike my own mind which has no beginning and no end, nor fixed measure of thought . . . other than its own cornerless mystery—to transcribe the latter in a form most nearly representing its actual occurrence." And lest we conclude that Olson and Ginsberg represent a particular Projectivist or Beat sensibility rather than the dominant poetic discourse of the period, consider Robert Lowell's explanation, in his *Paris Review* interview of 1961, of the turn toward free verse that characterized the *Life Studies* poems of 1959: "I couldn't," says Lowell, sounding not unlike Ginsberg, "get my experience into tight metrical forms. . . . I felt that the meter plastered difficulties and mannerisms on what I was trying to say to such an extent that it terribly hampered me."

Following this, the title of Robert Duncan's 1960 collection, *The Opening of the Field*, is invoked as an emblem of "the fidelity of verse form to the actual arc of feeling, the notion that rhythm, sound repetition, and lineation can enact the process of discovery."[37]

Perloff's construction of a common ground of poetic value shared by these figures accomplishes exactly what she intends: a canny assessment of a general tone in midcentury American poetics that has since come under suspicion in what she calls "the age of media." So if we are to distinguish Olson's poetics from the conceptual miasma Perloff specifies—a necessary task if we want to take seriously its relevance to the present—then we will need to draw some lines of demarcation between his concerns and those of other exemplars of "the natural look" cited by Perloff. This will involve further differentiating Olson's objectism from "organic form" and "free verse," while also clarifying the apparently problematic connection between objectism and Olson's breath poetics.

In a 1951 letter to Robert Creeley, Olson urges him to read the "GEOMETERS of the 19th century":

> FOR, it strikes me as crucial, it was not the physicists with their extraordinary analysis of the nature of the *substance* of matter, but the geometers who set out to figure the *disposition* of matter in space who opened up the kinetics we (& whitehead) are the inheritors of. Or perhaps it is better (these physicists were so valuable) to put it, that it was the initiative which was geometrical.[38]

For Olson the significance of non-Euclidean geometry is a theory of "*disposition* of matter in space," which implies that the formation of matter is involved with the formation of space, rather than merely occupying space. Matter and form cannot be considered independently, since the "substance" of materiality is indissociable from the formal disposition of actual materials. The principle at stake here is fundamental to our consideration of poetry as a branch of materials research and fabrication: the "nature" of material substance is contingent upon its configuration.

Olson gleans from Melville's *Moby Dick* the principle that "the inertial structure of the world is a real thing which not only exerts effects upon matter but in turn suffers such effects."[39] In other words, there is a retroactive relation between the disposition of matter and the field of relations in which matter is involved. Olson references

> [the] characteristic Riemannian observation, that the metrical structure of the world is so intimately connected to the inertial structure that the metrical field (art is measure)

> will of necessity become flexible (what we are finding out these days in painting writing and music) the moment the inertial field itself is flexible.[40]

Measure is posited here as a process immanent to material relationality, not as a detached quantitative standard. The practice of "composition by field" formulated by Olson in "Projective Verse" registers the metrical flexibility demanded by a model of relationality that is retroactive rather than unilateral, in which relations are as important as the materials they relate and determine. Considered this way, it is clear that what Olson calls "open field poetry" requires no appeal to an "open mind, open heart," as Perloff puts it. Rather, Olson's practice of composition by field follows from his effort to assimilate the counterintuitive data of post-Newtonian, post-Euclidean physical and metrical theory.

Following his reference to Riemann in the essay on Melville, Olson invokes Einstein as the physicist who established the flexibility of the inertial field "by the phenomena of gravitation, and the dependence of the field of inertia on matter." On the basis of these findings, Olson posits, we can state the following "absolute conditions":

> matter offers perils wider than man if he doesn't do what still today seems the hardest thing for him to do, outside of some art and science: to believe that things, and present ones, are the absolute conditions; but that they are so because the structures of the real are flexible, quanta do dissolve into vibrations, all does flow, and yet is there, to be made permanent, if the means are equal.[41]

Equal—as the title of the Melville essay has it—to the real itself. The "real" of Olson's title is the *composition* of materials through relational processes. His own practice of composition thus aspires to be equal to conditions of reality in which both the determinacy of matter and the fluidity of energy are understood as mutually constitutive, in which objects and relations can be both understood as flexible and made to hold. It is certainly *not* Olson's point that Riemannian geometry and post-classical mechanics have revealed the "life" of things. On the contrary, recognizing the energy and motion constitutive of matter makes it possible to recognize that "life" is not the necessary condition for a modality of being beyond inert fact.

Bearing these scientific concerns in mind, we can now return to Perloff's constellation of Levertov, Olson, Ginsberg, Lowell, and Duncan as progenitors of "the natural look," attempting to draw some lines of demarcation. Ginsberg states that conventional form is "too symmetrical, geometrical, numbered." On the basis of the Riemannian context in which Olson situates his poetics, one might demand of Ginsberg: to what kind of symmetry, geometry, or number do you refer? When Lowell claims, "I couldn't get my experience into tight metrical forms," we might pressure his apparent understanding of experience as the possession of a subject ("*I* couldn't get *my* experience") before suggesting that *no* experience can evade the constriction of a highly particular metrical form, though it need not be a standardized metric. For Olson, the inescapable rigor with which a particular body immanently specifies a metric is constitutive of an open field—of the density of a reality that is without interruption by the luxurious bracket that Ginsberg calls "my own mind." Again, for Olson, the discovery of the nineteenth century was that "quantity, the measurable and the numerable—was suddenly as shafted in, to any thing, as it was also, as had been obvious, the striking character of the external world, that all things do extend out."[42] That is: the measurable and the numerable cannot be wished away in favor of the special internal life of "my own mind" or "my experience;" rather, they inhere *within* any physical thing, within the body or the poem, and are expressed by its spatio-temporal extension.

In view of these distinctions, we can return to Olson's characterization of "Life" as "that dirty capital (doric, corinthian, OR iambic)." "To break the pentameter," writes Ezra Pound in a tribute to Walt Whitman, "that was the first heave."[43] It is in this sense that Whitman is the contemporary of Riemann. Olson links iambic pentameter to classicism ("doric, corinthian") insofar as it constitutes a predetermined frame in which the material activity of writing/speech unfolds. One might agree with Ginsberg that such a frame—which opens a "fixed" and predetermined space-time for activity—amounts to closure. But one could say the same of "my own experience" or "the cornerless mysteries" of "my own mind," which similarly carve out a territory of reality that belongs exclusively to "me." As Olson is evidently concerned to establish, one could say the same of "Life" insofar as it is determined as "an absolute," or otherwise than relative to conditions of reality that are "*ahead* of life." For Olson, both vitalism and humanism are classicist, as is iambic pentameter. In fact, the notion that iambic pentameter is the "natural" rhythm of English speech is itself construed by Olson as equivalent to the vitalist hypostatization of Life, or a humanist hypostatization of the subject: it implies that there is such a thing as a *generic* rhythm of English speech, which precedes and punctuates—which structures, or flows through or beneath—any particular rhythm. This is what Eliot called "the ghost of meter."[44] Insofar as it subordinates particularity to the abstract generality of a common currency, it operates (as Olson's pun implies) like capital.

The cases of Denise Levertov and Robert Duncan require similarly firm distinctions. Consider the central passage from Levertov's "Some Notes on Organic Form":

> A partial definition, then, of organic poetry might be that it is a method of perception, i.e., of recognizing what we perceive, and is based on an intuition of an order, a form beyond forms, in which forms partake, and of which man's creative works are analogies, resemblances, natural allegories.[45]

Levertov's intuition of "a form beyond forms, in which forms partake" has a great deal in common with Robert Duncan's poetics of "the dance," but it is antithetical to Olson's post-classical materialism, which breaks with any conception of an ideal form. "In the dawn that is nowhere," writes Duncan, "I have seen the willful children / clockwise and counter-clockwise turning."[46] And in the previous section of the same poem:

> I see always the under side turning,
> fumes that injure the tender landscape.
> From which up break
> lilac blossoms of courage in daily act
> striving to meet a natural measure.[47]

It is precisely such a "dawn that is nowhere" or a "natural measure" that one strives to "meet" to which Olson's materialist particularism is opposed. If Ginsberg and Lowell, in reacting against fixed meter, assert the independence of "my own mind" or "my experience," Levertov and Duncan invert this privative framing operation by projecting the Law of form beyond the particular world that it regulates. The opening of Duncan's sequence "The Structure of Rime" is representative of this tendency:

> I ask the unyielding Sentence that shows Itself forth in the
> language as I make it,
>
> Speak! For I name myself your master, who come to serve.
> Writing is first a search in obedience.
>
> . . .
>
> O Lasting Sentence,
> sentence after sentence I make in your image.[48]

For Olson, what is demanding about the flexible immanence of the real is the specificity of its occasions, and there is no "unyielding sentence" to whose "image" any *particular* sentence might aspire to be adequate.

In "Projective Verse," Olson transcribes an aphorism that he gleans from Creeley: "FORM IS NEVER MORE THAN AN EXTENSION OF CONTENT."[49] In "Some Notes on Organic Form," Levertov adds a corollary to Creeley's formulation: "form is never more than a *revelation* of content."[50] This is among the characteristic gestures of Romantic organicism: to convert the physicality of "extension" into a "revelation," seeking an ideal form behind or above the physicality of the world. It is also the operative principle behind that movement of Ezra Pound's *Cantos* that Hugh Kenner has demonstrated so thoroughly:[51] their oscillation between a reverence for "the green world"[52] and a neo-Platonic idolatry of "Lux in diafana / Creatrix."[53] For Pound, both the organic and the celestial are models of "scaled invention or true artistry"[54] for the poetic creation of a "paradiso / terrestre."[55] The adequation of the "green casque"[56] of Pound's *Cantos* to Duncan's "dawn that is nowhere" is precisely the "natural measure" that organic form strives to meet through an idealist mimesis. Pound's emblem of a true artistry that *participates* in an ideal form is the "room in Poitiers where one can stand / casting no shadow."[57] In such an ideal room, the confluence of built structure and the light flowing through it constitutes the perfect correspondence between a material body and a form beyond forms in which the room partakes, a correspondence into which any "shadow"—any index of residual physicality—vanishes.

In "Projective Verse," Olson states that a poem should be a "high energy-construct."[58] Alongside this demand, one should also bear in mind the value Olson finds in what he calls "the difficulties" of "the simplest things."[59] What these "difficulties" have in common with a "high energy-construct" is that they register the constitutive excess and activity agitating any apparently fixed material structure. But Olson's affirmation of "the difficulties" recognizes such excess in the formal irregularities and "deficiencies" that undermine structural perfection, necessitating a somewhat humbler mode of what Pound calls "scaled invention or true artistry":

> I count the blessings, the leak in the faucet
> which makes of the sink time
>
> . . .
>
> Or the plumbing,
> that it doesn't work, this I like, have even used paper clips

as well as string to hold the ball up And flush it
with my hand
. . .
Holes
in my shoes, that's all right, my fly
gaping, me out
at the elbows, the blessing
that difficulties are once more
. . .
In the land of plenty, have
nothing to do with it
take the way of
the lowest,
including
your legs, go
contrary, go

sing[60]

"The leak in the faucet / which makes of the sink time": it is this situated commitment to contingent occasions of material fallibility that is the enduring value of Olson's localism, more so than the apparently larger project of mythopoetic redemption at issue in the *Maximus Poems*. The fact that he does not seek to eliminate "the difficulties" he records is also the essential difference between Olson's sensibility and the ideology of efficient form characteristic of technoscience at the limits of fabrication, whether the mid-twentieth-century cybernetics in which Olson was steeped or twenty-first-century materials science.

The problem with Levertov's corollary to Creeley's maxim is that it elides the significance of "extension" as a *geometrical* term. Creeley is concerned with the spatio-temporal disposition of matter. "Form IS the extension of content," he writes to Cid Corman.

> The use, USE, by any means whatever, anything that will work, of MATERIAL (sky man horse, steel iron coal, potato onion carrot, H20 analysis fission) MATERIAL (OBJECT), to make known, to DECLARE: CONTENT. Fused, driven thru, beaten, hammered, fired: FUSED—IS FORM, is the FORM. No "patterns", nothing but what is SO MADE.[61]

This does not sound particularly "organic" or "natural," nor is it compatible with any aspiration toward "a form beyond forms." "No 'patterns,'" Creeley insists. He states that form *declares* content, makes it known. But what Levertov would describe as a "revelation," Creeley renders as a chiasmatic nexus of materials through which "inside" and "outside" undergo an exchange:

> the materials of poetry or of prose (materials: the externals, the phenomena outside the given man, around him, OBJECTS) are that which the content IN man uses to declare itself—that this declaration, this COMING OUT by means of the materials, MAKES the form—so it might go, that being my own thought as to HOW form discovers itself.[62]

The key insistence here is that the extension of content occurs "by means of the materials." Form "discovers itself" not through revelation but rather fabrication. Form is "fused," in and as material extension.

Olson is even clearer than Creeley in dismissing any ideal form or subject from the poetics of objectism. "I am not able to satisfy myself," he writes,

> that these so-called inner things are so separable from the objects, persons, events which are the content of them. . . . What I do see is that each man does make his own special selection from the phenomenal field and it is true that we begin to speak of personality, however I remain unaware that this particular act of individuation is peculiar to man, observable as it is in individuals of other species of nature's making (it behooves man now not to separate himself too jauntily from any of nature's creatures).[63]

Individuation, for Olson, is a matter of "selection from the phenomenal field," a matter of prehensions, to use Whitehead's vocabulary. As such, it is not "peculiar to man," since any actual entity or occasion is determined through prehensive relations. It behooves us, in Olson's estimation, not to essentialize our modality of being-in-the-world so as to divorce it from that of other creatures. But his appeal to "nature" in this passage is not an appeal to "the natural look." An effort to situate "man" within a common world does not involve, in Olson's case, a facile appeal to spontaneous freedom, to organic form, or to an absence of metrical regulation. On the contrary, we have seen that Olson's understanding of nature is grounded in scientific theories that enable him to think *physicality* in a manner at odds with any defense of "my own mind" or "my experience" against "the measurable and the numerable," at odds with the transcendence of any "form beyond forms," and at odds with any poetics of "revelation."

If we have thus distinguished Olson from the exemplars of "the natural look" enumerated by Perloff, we still need to grapple with the relation between Olson's objectism and his theory of breath poetics. What is the significance, for objectism, of the claim that "the line comes (I swear it) from the breath, from the breathing of the man who writes, at the moment that he writes?"[64]

As Steve McCaffery has astutely noted, we cannot simply attribute such a claim to phonocentrism, since "breath and graphism are brought into an interesting complicity" by Olson's formulation.[65] Further, "the breathing of the man who writes" is crucially mediated, for Olson, by the technical apparatus of the typewriter, which enables the scoring of the written and published poem according to the rhythm of breath through the rigidity and precision of its spacing, the reproducibility of the typewritten inscriptions. The deconstructive critique of phonocentrism challenges the binary distribution of speech and writing and the metaphysical priority of the former term, whereas Olson insists upon the simultaneous complicity of orality and graphism within a technically mediated and reproducible inscription.

Nevertheless, what are we to make of Olson's other claims regarding speech in "Projective Verse"—such as, for example, "breath allows *all* the speech-force of language back in (speech is the 'solid' of verse, is the secret of a poem's energy)"?[66] Here we should be as exacting as we were with Ginsberg or Lowell: to what kind of "solid" does Olson refer? Olson's stance toward reality is to address all "solids" in terms of processes of interaction traversing myriad relations and scale levels. As does Whitehead, he affirms the counterintuitive findings of post-classical science with respect to the distributed relationality of objects that are apparently confined to a simple location. For Olson, "speech is the 'solid' of verse, is the secret of a poem's energy" not because it anchors the iterability of writing in the metaphysical presence of a subject but because breath—the operation of the respiratory system—manifests the non-coincidence of the body to itself, its participation in an outside, and the registration of respiratory rhythms in the writing of the poem inscribes this outside within its own parameters. Speech is at once the "solid" of verse and the secret of a poem's energy insofar as it draws both the body and the poem into the field of the open. It is because Olson approaches physiology and language—the physiology *of* language—not only in terms of biology or linguistics but also in terms of physics, geometry, and information theory that his poetics is not "organic."

One thus has to grapple with the full complexity of Olson's thinking of the solid:

> because, now, a poem has, by speech, solidity, everything in it can now be treated as solids, objects, things; and, though insisting upon the absolute difference of the reality of verse from that other dispersed and distributed thing, yet each of these elements of a poem can be allowed to have the play of their separate energies and can be allowed, once the poem is well composed, to keep, as those other objects do, their proper confusions.[67]

Note that this closing emphasis upon retaining "the proper confusions" of objects inverts the final emphasis upon "permanence" in the last sentence of "Equal, That Is, to the Real Itself" ("the structures of the real are flexible, quanta do dissolve into vibrations, all does flow, and yet is there, to be made permanent, if the means are equal"). The separate energies of the poem are to be made permanent as a field of tensions *by* retaining their proper confusions, "as those other objects do." This emphasis upon the co-implication of permanence and proper confusions demands that we bear in mind Whitehead's difficult precept: that while "the notion of a 'common world' must find its exemplification in the constitution of each actual entity, taken by itself for analysis," that analysis has to be undertaken in full cognizance of the fact that "every item of the universe, including all the other actual entities, is a constituent in the constitution of any one actual entity."[68] The boundedness of a poem or region of space—the field of difference it "contains"—depends for its boundedness upon its situation in a larger field of differences.

This mutual determination of objectism and open field poetics is stated clearly enough by Olson:

> The objects which occur at every given moment of composition (of recognition, we can call it) are, can be, must be treated exactly as they do occur therein and not by any ideas or preconceptions from outside the poem, must be handled as a series of objects in field in such a way that a series of tensions (which they also are) are made to *hold*, and to hold exactly inside the content and the context of the poem which has forced itself, through the poet and them, into being.[69]

An object *is* a field of tensions, and the poem consists of a series of objects in field that must be handled such that a series of tensions (objects) are made to hold, are made to cohere, are *composed*. But how is this practice of composition by field capable of drawing together "the breathing of the man who writes" with the assertion that "man is himself an object"?

The dynamics of the relation between open field poetry, projective verse, and objectism depend primarily upon Olson's complicated determination of the body's proprioceptive operation, to which we must return:

> PROPRIOCEPTION: the data of depth sensibility/the "body" of us as
> object which spontaneously or of its own order
> produces experience of, "depth" Viz
> SENSIBILITY WITHIN THE ORGANISM
> BY MOVEMENT OF ITS OWN TISSUES[70]

The body is understood as both an organism and an object, and when Olson insists that "man is himself an object" he emphasizes the primacy of the latter term. For Olson, the body's experience as an organism is an aspect of its existence "as object." Again: *the body as organism is a subset of the body as object.* The body's network of discrete, relational components is not subsumed by and subordinated to a systemic organization imposing a sharp distinction between subject and object, organism and environment. "Kinesthesia" is defined by Olson as "the sense whose end organs lie in the muscles, tendons, joints, and are stimulated by bodily tensions (—or relaxations of same)."[71] The proprioceptive sense of the body's movement thus recapitulates the "character of being," physicality, of both the object and the poem: it is composed of a series of tensions that hold together not as an overdetermining system but *through* and *as* a field of relational components. A body is a collective rather than a system, an object constituted of and among objects, such that "the skin itself," the porous surface at the interface of "interior" and "exterior" components, "is where all that matters does happen." Along this surface, or *at* this interface, "man and external reality are so involved with one another that, for man's purposes, they had better be taken as one."[72]

Breath may be what flows across this interface, but it is not therefore the substance of a *metaphysical* union. Olson states that "breath is man's special qualification as an animal."[73] This is a rather peculiar determination, but it is not exactly essentialist. It is a distinction at the level of species (phylogeny), not at the level of being (ontology). One might say: buzz is bee's special qualification as an animal, or 28.0855 is silicon's special qualification as an element. These statements do not involve claims about what an entity essentially *is*. Like Olson's, they are statements concerning special "qualifications" appropriate to particular rhetorical contexts. The breathing of a human body, which can be measured by verse insofar as it is involved in both speech *and* writing, is a particular way in which "physicality" is qualified. The respiratory apparatus participates in and enacts the distribution of physicality through its passage across the threshold of a body, just as electrons enact the distribution of physicality when they flow across the interface between the tip of a Scanning Tunneling Microscope and the surface to which it is proximate.[74] Both are generative of measure. And both of these instances illustrate the principle that there is no immediate action at a distance: that a continuum of physical interaction operates as a field. "Only to space-time coincidence and immediate space-time proximity can we assign an intuitively evident meaning," Olson writes in *The Special View of History*.[75]

Any "voice" that ever has anything to do with a poem—whether of its writer or reader—is always a particular, proximate voice (Whitehead would say an "actual occasion"), though it is never really "my" voice. It is not a metaphysical substance or a mystical power. Breath is *specific* to the energy of a particular body's inherence within space-time, just as the frequency of a particular molecule's vibration is specific to its chemical structure, or as the wave function of an electron is specific to its behavior. These are not idle analogies, since for Olson it is Heisenberg's uncertainty principle that reasserts the value of Keatsian negative capability, the capacity to persist in doubts and uncertainties. A breath line is "specific" in the same sense that a wave function is specific: it is produced by, and in that sense includes, *all* the force of the field of tensions involved. But that does not mean that the entirety of that field of force can be "positively identified and detained," as Thomas Pynchon might say.[76] In other words, what Olson calls "force" is not necessarily a matter of *presence*. For Olson, speech is the "solid" of verse like an electron is the "solid" of matter or like the arche-trace is the "solid" of *écriture*. It is a ~~solid~~, which is no doubt why Olson embeds the word in scare quotes. It does not fix the simple location of a body/object/event; it is the secret of its energy.

We are now in a position to make sense of the context in which Olson refers to "secrets objects share":

> It comes to this: the use of a man, by himself and thus by others, lies in how he conceives his relation to nature, that force to which he owes his somewhat small existence. If he sprawl, he shall find little to sing but himself, and shall sing, nature has such paradoxical ways, by way of artificial forms outside himself. But if he stays inside himself, if he is contained within his nature as he is participant in the larger force, he will be able to listen, and his hearing through himself will give him secrets objects share. And by an inverse law his shapes will make their own way. It is in this sense that the projective act, which is the artist's act in the larger field of objects, leads to dimensions larger than the man.[77]

This "inverse law" of Olson's poetics, that relative containment is a prerequisite of participation, and vice versa, is the suture that threads together objectism, open field poetry, and projective verse. It formalizes, in the context of Olson's poetic practice, the retroactive relation of metrical and inertial fields that Olson finds central to post-classical science. Breath poetics does not involve an essentialism of the organic body that is contrary to objectism. Rather, it is the system of calibration by which Olson attempts to remain at once "inside himself" and participant in "the larger field of objects." This opening of a common world through the particular gradations of relevance determinate of a specific physicality is the open secret that objects share.

Olson's breath poetics emerges from upheavals in our understanding of natural law, but his poetics is indifferent to the cultivation of "the natural look." The geometries and physical theories discovered between 1850 and 1950, upon which Olson bases his thinking about the body and the poem, are thoroughly counterintuitive and decidedly "unnatural" by a commonsense standard valuing whatever *looks* natural. Olson's poems do not mimic the appearance of phase-space diagrams, but their formal properties result from a rigorous effort to index an object that isn't a simple location rather than from an assertion of self-evident presence, a careless disregard for measurable quantity, or a desire for "personal" freedom.

"Free verse = freedom; open form = open mind, open heart": the mid-century equations that Perloff cites in *Radical Artifice* are, as she points out, facile. To the extent that they were accepted as axiomatic, they are rightly to be abhorred. But such equations are equally facile if we invert their normative value: free verse = self-indulgence; open form = foggy mysticism. Thus if we want to think both capaciously and accurately about Olson's place in literary history, we need to be at once more specific and less restrictive in our mapping of poetic common ground.

When Craig Dworkin splices linguistics and geology textbooks to create a treatise on "tectonic grammar" in *Strand*; when Christian Bök "misreads the language of poetics through the conceits of geology" in *Crystallography*; when Tan Lin writes in *BlipSoak01* that the most beautiful poem "would resemble non-designed furniture . . . or biofeedback devices," and that "every moment deserves extension via something"; when Caroline Bergvall records the textual distortions resulting from a dental extraction prior to a reading of her poem "About Face," exploring a "choreography of the physical mouth into language"[78]—when we consider any of these instances of contemporary writing, we should recognize that these poets are writing in the tradition of Olson's work just as surely as that of John Cage, Steve McCaffery, Andy Warhol, Marina Abramovic, or Vito Acconci—even if they wouldn't be quick to cite Olson's objectism as a precedent. Recognizing *that* shared secret—a hybrid tradition of objectism, proceduralism, conceptualism, and performance poetry—would give us an entirely new way to think about what we mean by "Black Mountain poetry," a tradition that should include the procedural and conceptualist poetics of John Cage as well as the objectism of Olson and Creeley.

OBJECTISM VERSUS OBJECTIVISM

Olson introduces the term "objectism" to distinguish his stance toward reality from that of "objectivism," a movement initiated by a special issue of *Poetry* magazine in 1931, edited by Louis Zukofsky and including contributions by George Oppen, Charles Reznikoff, Carl Rakosi, and Basil Bunting. As Olson notes, objectivism was closely aligned and affiliated with the poetics of his major predecessors, Ezra Pound and William Carlos Williams:

> It is no accident that Pound and Williams both were involved variously in a movement which got called "objectivism." But the word was then used in some sort of a necessary quarrel, I take it, with "subjectivism." It is now too late to be bothered with the latter. It has excellently done itself to death, even though we are all caught in its dying. What seems to me a more valid formulation for present use is "objectism."[79]

While this is a passage that has received a great deal of attention from Olson's critics, the stakes of Olson's distinction need to be grasped more clearly. In *Enlarging the Temple*, Charles Altieri quotes this passage but does not analyze the distinction between objectism and objectivism.[80] In an attack on what she considers the entirely derivative character of Olson's poetics, Marjorie Perloff dismisses Olson's distinction as "irrelevant."[81] In his chapter on Olson, Whitehead, and the objectivists, Robert von Hallberg argues that Olson's position "was much closer to the Objectivists than he admitted."[82] In reply, one might note that Olson does not insist too much on his distance from the objectivists; rather, his rhetoric in this case is uncharacteristically tentative and measured. Olson acknowledges that pitting "objectivism" against subjectivism was "a necessary quarrel"; he merely suggests "what seems to me a more valid formulation for present use." But the larger problem with von Hallberg's account is that he misses the significance of the modest distinction upon which Olson *is* trying to insist. "In 1950," von Hallberg writes, "Olson was arguing, as we have seen, for an epistemological shift toward what he called 'objectism.'"[83] What von Hallberg thus elides is Olson's emphasis upon an *ontological* rather than epistemological approach to a poetics of the object. It is this point of emphasis that constitutes the key difference between objectism and objectivist poetics.

We can determine precisely what objectivist and objectist poetics have in common, and where they differ, by examining a passage in Whitehead's *Science and the Modern World*—a passage that Olson may have in mind when he remarks on objectivism's "necessary quarrel" with "subjectivism." "The objectivist," Whitehead writes, "holds that the things experienced and the cognisant subject enter into the common world on equal terms" (*SMW* 89).[84] This is a principle in which Olson, the objectivists, and Whitehead seem equally invested. But Olson implies that in its necessary quarrel with subjectivism (the attribution of a primary role to the cognizant subject), objectivist poetics stresses the *epistemological* implications of considering the cognizant subject and the things experienced on equal terms. These epistemological implications of objectivism are elucidated by Whitehead as follows: "The things experienced are to be distinguished from our knowledge of them. So far as there is dependence, the *things* pave the way for the *cognition*, rather than *vice versa*" (*SMW* 88–89, italics in original). This is the epistemological position of modern science, a position that Olson and Whitehead share, as far as it goes. But what distinguishes Olson's or Whitehead's approach to "physicality" is the emphasis they place upon the ontological implications of this epistemological position. Olson follows Whitehead by stressing that "we seem to *be* ourselves elements of this world in the same sense as are the other things which we perceive" (*SMW* 89, my italics). In this statement, the "cognising subject" who enters into the common world on "equal terms" with "the things experienced" becomes an *element* of this world *in the same sense as those things*. This is a subtle but decisive difference, insofar as the former proposition is concerned with the *cognition* of objects by a subject, while the latter is concerned with the modality of *being* of any element of the world whatever. When Olson proposes that "man is himself an object," he makes a claim that is ontological rather than epistemological.

Louis Zukofsky wrote this compelling sentence twenty years before "Projective Verse" was published: "*Impossible* to communicate anything but particulars—historic and contemporary—things, human beings as things their instrumentalities of capillaries and veins binding up and bound up with events and contingencies."[85] Here Zukofsky emphasizes his understanding of "human beings as things" in a way that is consistent with Olson's or Whitehead's thought—but to the extent that this position matters to Zukofsky, Olson is probably right to suggest that objectism would be a preferable term. There is no need to insist upon any absolute break between Olson and his mentors, only to recognize that the primary emphasis of objectivist poetics rests upon the objective perception of things rather than upon the claim that "man is himself an object." It is no disparagement of Zukofsky or Oppen to point out that their poetic concerns are not so pervasively saturated by the ontological problem of the object in wake of post-classical physics as is Olson's. Oppen writes,

Not to reduce the thing to nothing—

I might at the top of my ability stand at a window
and say, look out; out there is the world.[86]

Above all, Oppen values "simply pointing to things—and clearly enough or accurately enough."[87] Olson is equally committed to what Oppen calls "a limited, limiting clarity":[88]

snow
coming
up to my window, going up
and as well across
in front of my glasses in front of my eyes and
two thicknesses of glass windows also
between me and
the outside, trying to see the actual flakes
(*M* III.179)

The particular objects of the field in which the body feels itself situated are never reduced to nothing. But in the process of "trying to see the actual flakes," Olson is also pervasively concerned with what "the actual flakes" *are*. In the *Maximus* poems, the objects of experience—whether stellar dendrites, mythological narratives, the constellation of perspectives that compose a city, or "elements (of glacial / accumulations" (*M* II.148)—are taken up by Olson within a context that cannot be "simply" pointed to:

Watch-house
Point: to descry
anew: *attendeo*
& broadcast
the world (over the
marshes to the outer limits even when minutiae
hold & swim in the electro-magnetic
strain
(*M* III.25)

The *Maximus Poems* are replete with resolute particulars, as is Oppen's *Of Being Numerous*, and one would not want to trade these for anything ("the motes / In the air, the dust // Here still").[89] But in Olson's *Maximus Poems*, the condition of the particulars with which the poem is composed cannot evade the radical desubstantializations to which objects are subject at "the outer limits" of physical being, "in the electro-magnetic / strain." As Zukofsky notes, it is "*impossible* to communicate anything but particulars": no matter how "general," "abstract," or "universal" a proposition, anything that is communicated thereby *becomes* particular under the conditions of its utterance or reception. Nonetheless, we cannot ignore the strange testimony of the physical sciences that while these numerous particulars are "here still," it is nevertheless true that "in a certain sense everything is everywhere at all times. For every location involves an aspect of itself in every other location" (*SMW* 91). Whitehead, Olson, and Zukofsky could perhaps agree on this double determination of a particular context. In the final lines of Zukofsky's lifework, *"A,"* we encounter the following convocation:

> —impossible's
> sort-of think cramp work x: moonwort:
> music, thought, drama, story, poem
> parks' sunburst—animals, grace notes—
> z-sited path are but us.[90]

Here, under the sign of "us," Zukofsky draws together a common world composed of texts, ideas, events, sounds, and creatures along a "z-sited path." Acknowledging the common ground his project shares with Olson's, we can also acknowledge that Olson extensively theorizes what Zukofsky only briefly alludes to: the operations of the body as an object within the field of the poem within the field of reality, "this extended skin of our own / composite body—miracle of / form" (*M* 561).

In the intensity of his engagement with the counterintuitive "character of being" he calls "physicality," Olson is perhaps more proximate to Ezra Pound's figures of the poem as a vortex, a radiant node, or a radiogram than he is to Oppen or Zukofsky.[91] But Olson extrapolates from what Daniel Albright calls Pound's "quantum poetics"[92] by including "the body of us as object" within the field of the poem. In doing so, Olson insists on "the open field" of the body and the poem through an ontology that destabilizes objectivist epistemology: the critique of "simple location" in both Whitehead and Olson is sufficiently radical to discredit the prospect of "simply pointing to things" that motivates Oppen's poetics of clarity. As we have seen, Olson integrates the "proper confusions" of objects in a manner that extends the scale of our attention to the conditions of their coherence from the minimal units of "the actual flakes" to the "outer limits" of "the electro-magnetic / strain." In concert with Whitehead's philosophy, Olson gives us a poetics of the object-event, of the actual entity/actual occasion. He goes to tremendous lengths to at once *specify* the event that his physical existence as an object *is* and to include that particular object-event within the larger field of "those other creations of nature which we may, with no derogation, call objects."[93] We lose what matters about this effort if we do not distinguish objectism from objectivism while also acknowledging their proximity.

OBJECTISM VERSUS AUTOPOIESIS

In *The Poetics of the Common Knowledge*, Don Byrd attends in some detail to the relevance of cybernetics and information theory to Olson's poetics. While Olson was directly influenced by the cybernetic theory of Norbert Wiener, Byrd argues that his poetics anticipates the "second-wave" systems theory formulated by Humberto Maturana and Francisco Varela in their paper "Autopoiesis and Cognition: The Organization of the Living" (1973), unpublished in English until 1980, ten years after Olson's death.[94] The theory of autopoiesis (eventually elaborated as a theory of social systems by Niklas Luhmann and now widely influential in literary and media/technology studies)[95] offers an account of recursive, self-organizing processes by which living systems are differentiated from their environment as operationally closed unities capable of reproducing their organization. What crucially distinguishes autopoietic theory from Wiener's first-wave cybernetics is the thesis of operational closure: whereas Wiener's cybernetics understood a system as an input-output mechanism, regulating itself through feedback, autopoietic systems "do not have inputs or outputs"; they are defined by their self-reflexive "autonomy" from their environment.[96] In *The Human Use of Human Beings*, Wiener argued that

> we, as human beings, are not isolated systems. We take in food, which generates energy, from the outside, and are, as a result, parts of that larger world which contains those sources of our vitality. But even more important is the fact that we take in information through our sense organs, and we act on the information received.[97]

By contrast, autopoietic systems do not "take in information." Their self-reflexive operations are "perturbed" by an environment to which they are "structurally coupled," but no information crosses the organizational boundary distinguishing system from environment, within which information is internally and autonomously constructed. Maturana and Varela distinguish autopoietic or living systems, which "are autonomous; that is, they subordinate all changes to the maintenance of their own organization," from allopoietic systems, which "have as their product something different from themselves."[98]

Byrd values the emphasis of autopoietic theory upon homeostatic systematicity as a property of life, and he values the resources the theory offers for thinking operations of non-logical (i.e., operational) distinction as prerequisite to securing "a space in which it is possible to articulate an increasing complexity of self-organization."[99] Concerned to establish the self-organizing independence of Olson's work, and the mode of knowledge it makes available, from what he calls the "statistical reality" of the social realm, Byrd wants to claim that Olson's work is concerned with "nonstatistical action" of "separate biological organism[s]" as "radically autonomous beings." "Nonstatistical action," Byrd writes,

> arises from entities that are not preconditioned by their membership in a society. Unlike conventional atoms that arise in our efforts to use a general system of signs for particular objects, whether irreducible units of radium as studied by physicists or units of an electorate studied by public-opinion experts, the human "atom" is not social—which is to say, not an atom.[100]

"In nature," argues Byrd, "everything is becoming something else: everything is a transfer of energy or information, a machine that produces something other than itself from itself," whereas human beings "draw themselves forth from the human . . . they do not act to produce something other than what they are; they act only to maintain their own identity."[101] As a means to the end of articulating this vision of human autonomy through Olson's work, Byrd argues that Olson gradually turns away from the model of feedback based on input-output systems he had gleaned from Wiener's cybernetics toward a model of autonomous self-organization that anticipates autopoietic theory.

Autopoietic theory is itself at odds with Byrd's radical distinction of "the human" from "nature" since it defines all living systems as autopoietic. But in either case, foregrounding the importance of objectism as Olson's "stance toward reality" helps us make clear both why Byrd's account is untenable and why Olson's poetics are incompatible with autopoietic theory. We recall that Olson insists upon the importance of situating "reality" *ahead of* "life," such that the latter must be understood as strictly relative to the latter. But we recall also Olson's claim that the relation between "man" and "nature" is such that "if he stays inside himself, if he is contained within his nature as he is participant in the larger force, he will be able to listen, and his hearing through himself will give him secrets objects share."[102] This "inverse law" brings us closer to the primacy of operational closure constitutive of autopoietic systems. The difference, however, is that the operationally closed unities of autopoietic theory cannot partake of "secrets objects share" because their capacity to "stay inside" themselves through their circular recursivity is precisely what separates them as living systems from non-living objects. An autopoietic system does not act "in the larger field of objects" because it only behaves in the domain of its own circular processes; "its phenomenology is the phenomenology of a closed system."[103] The operations of autopoietic systems do not "lead to dimensions larger than man"[104] because, as Byrd correctly insists, they "act only to maintain their own identity." For autopoietic theory or for Byrd's *Poetics of the Common Knowledge*, to recognize that "man is himself an object" is merely to capitulate to an allopoietic function in what Byrd terms "statistical reality." While Olson celebrates Whitehead for "getting the universe in" (*M* II.79), and while he recognizes post-classical geometry and physics for resituating man as "a thing among things" in a reality "without interruption," *Autopoiesis and Cognition* begins by stipulating that "a universe comes into being when a space is severed into two."[105]

"Some people," writes Whitehead, "express themselves as though bodies, brains, and nerves were the only real things in an entirely imaginary world. In other words, they treat bodies on objectivist principles, and the rest of the world on subjectivist principles" (*SMW* 88). This is what autopoietic theory does, using data from experiments on nervous systems and visual cortexes to construct a theory of phenomenal self-reflexivity. Whitehead insists that "this will not do; especially when we remember that it is the experimenter's perception of another person's body which is in question as evidence" (*SMW* 91). As opposed to a scientistic approach to the organism contradicted by subjectivist phenomenology, or Byrd's separation of biological organisms from the social, Olson follows Whitehead in thinking the co-constitution and mutual interdependency of objects and societies. Consider his attention to the following passage in his copy of *Process and Reality*, marked from multiple readings in pencil and in two different shades of ink (Fig. 11).

The "special societies" to which Whitehead refers are described as vehicles of order through which individual occasions realize "peculiar 'intensities' of experience," and of which the physical world exhibits "a bewildering complexity . . . favouring each other, competing with each other." "Source of art forms," Olson writes in the margin of this passage. He underlines each of the following, separately: "*regular trains of waves*, *individual electrons*, *protons*, *individual molecules*, *societies of molecules* such as *inorganic bodies*, *living cells*, and *societies of cells such as vegetable and animal bodies*." There is nothing here that acts only to maintain its own identity. Rather, the common world is composed of a multiplicity of collective societies, constantly participating in processes of con-

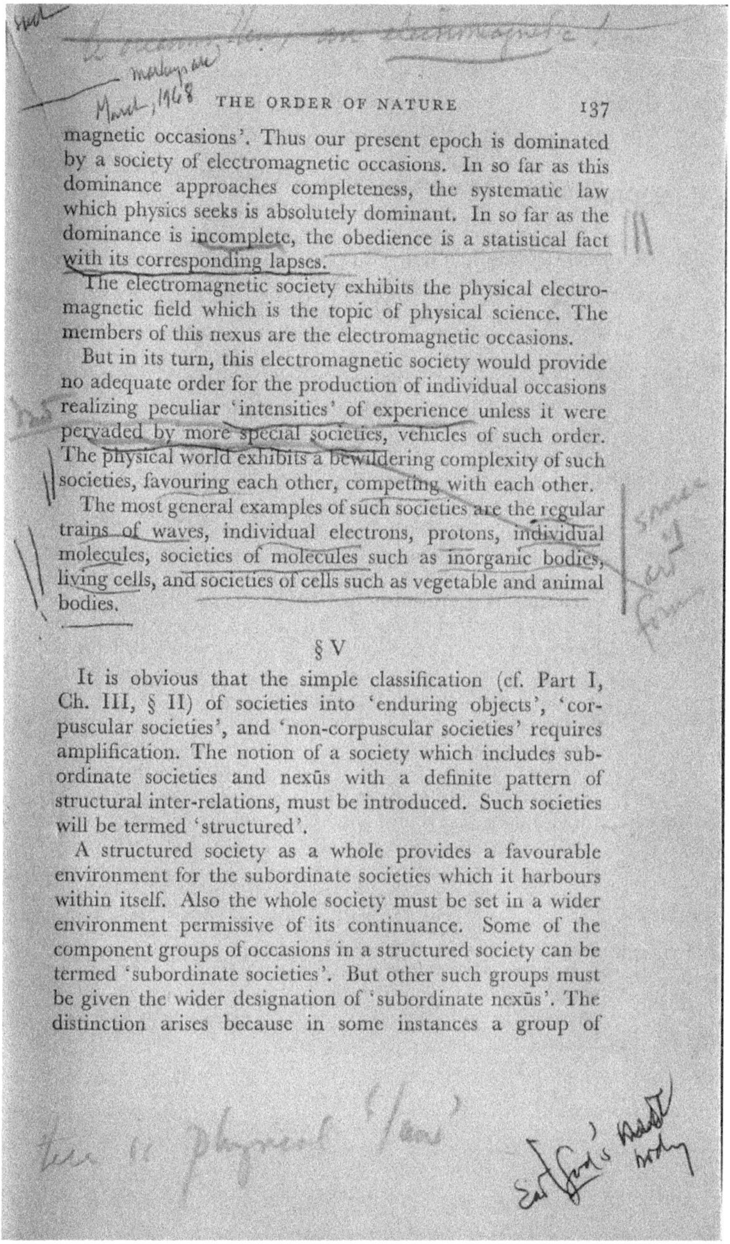

THE ORDER OF NATURE 137

magnetic occasions'. Thus our present epoch is dominated by a society of electromagnetic occasions. In so far as this dominance approaches completeness, the systematic law which physics seeks is absolutely dominant. In so far as the dominance is incomplete, the obedience is a statistical fact with its corresponding lapses.

The electromagnetic society exhibits the physical electromagnetic field which is the topic of physical science. The members of this nexus are the electromagnetic occasions.

But in its turn, this electromagnetic society would provide no adequate order for the production of individual occasions realizing peculiar 'intensities' of experience unless it were pervaded by more special societies, vehicles of such order. The physical world exhibits a bewildering complexity of such societies, favouring each other, competing with each other.

The most general examples of such societies are the regular trains of waves, individual electrons, protons, individual molecules, societies of molecules such as inorganic bodies, living cells, and societies of cells such as vegetable and animal bodies.

§ V

It is obvious that the simple classification (cf. Part I, Ch. III, § II) of societies into 'enduring objects', 'corpuscular societies', and 'non-corpuscular societies' requires amplification. The notion of a society which includes subordinate societies and nexūs with a definite pattern of structural inter-relations, must be introduced. Such societies will be termed 'structured'.

A structured society as a whole provides a favourable environment for the subordinate societies which it harbours within itself. Also the whole society must be set in a wider environment permissive of its continuance. Some of the component groups of occasions in a structured society can be termed 'subordinate societies'. But other such groups must be given the wider designation of 'subordinate nexūs'. The distinction arises because in some instances a group of

Figure 11 Annotations in Charles Olson's copy of Alfred North Whitehead's *Process and Reality*. Courtesy of Thomas J. Dodd Research Center, University of Connecticut.

stitution and differentiation. These collectives are at once the "special societies" that we are and the common world we compose. That is what "objectism" means.

THE CONGERY OF PARTICLES

In *The Sense of the World*, Jean-Luc Nancy speaks of "being a 'fragment' of a world whose matter is the very fraying [*frayage*] or fractality of fragments, places, and takings-place."[106] The matter of the final volume of Olson's *Maximus Poems* is composed of such scraps and fragments, places, and takings-place, teeming with the bewildering complexity of the sense of the world continually reiterating itself through societies on all scales, and then dissolving into distances. "Snow / coming / to my window," Olson writes. And then: "ah now no snow at all" (*M* III.179). The anagrammatic play of this last phrase minutely indexes the fact that Olson's later work is a fractured catalog of "matter forming itself, form making itself firm"[107] and then flickering into

light signals & mass points
normal mappings of
inertia & every possible action

of aether and of

change
(*M* III.133)

The body/object/event is *exposed* to this mutable sense of the world, and it is bound together in its fragility by that world's network of relations:

a tied
ellipse in which a story
is told and the ends of the strings are
laid over and stand
(*M* III.210)

The poem is at once structured and flexible, and it is replete with contingent unities, tied together, such as "Mother Earth Alone" (*M* III.226). But in the *Maximus* the earth itself is also a fragmented totality dispersing itself into "Ice-age megaliths" (*M* III.228), "high-lying benches / of drift material" (*M* III.16), "Darkness of the stones tumbled / in their own congeries" (*M* III.122), "the sea's // boiling the land's // boiling all the winds" (*M* III.186), and

the rocks
melting
into the sea, the forests,
behind, transparent
from the light snow showing
lost rocks and hills
which one doesn't, ordinarily,
know, all the sea
calm and waiting, having
come so far
(*M* III.107)

The import of such passages is that any unity is a *composition*, a social occasion constituted not only by a relational collective of diverse entities but by the entire history of their being together, "having / come so far." Olson's poetry is consonant with Whitehead's recognition that "everything perceived is in nature. We may not pick and choose."

> For us the red glow of the sunset should be as much a part of nature as are the molecules and electric waves by which men of science would explain the phenomena. . . . We are instinctively willing to believe that by due attention, more can be found in nature than that which is observed at first sight, but we will not be content with less. . . . The real question is, When red is found in nature, what else is found there also? Namely, we are asking for an analysis of the accompaniments in nature of the discovery of red in nature.[108]

The *Maximus Poems* might be parsed as a massive catalog of such "accompaniments," synthesizing an analysis of "history" and of "nature" by constantly looking for "what else" can be found in any given occasion. The poet happens upon perceptions and pursues their accompaniments:

the moon once more
in its piece missing
free in the sky & the harbor
all free   and orange too
as with no more reason than
thirst or another desire I
with out wish & full of
love, leave it to
itself
& tell you I have——?

It is actually
the red light
on the harborside
of the bridge
which made it glow so
from upriver
as I come nearer
to end
this night
(*M* III.159)

The orange glow of the harbor "is actually" due to "the red light / on the harborside / of the bridge" (rather than due to the moon?). This discovery is not a reduction of the poetic function of pathetic fallacy ("thirst or another desire") to empirical explanation but an expansion of the field of experience by the analysis of an accompaniment, of *something else*, of *more* in nature (the moon, the harbor, *and* "the red light / on the harborside / of the bridge") that comes into relation with an object of perception (an orange glow). The "I" of such passages is exhaustively curious, but it is not so much a subject pointing *at* objects as one of many body/object/events affected *by* other objects:

Full moon [staring out window, 5:30 AM March 4th
1969] staring in window one-eyed white round clear
giant eyed snow-mound staring down on snow-
covered full blizzarded earth
(*M* III.206)

Sun
right in my eye
4 PM December 2nd arrived
at my kitchen
window blazing
at me full in the
face approaching
the hill it sets
behind glaring
in its burst of late
heat right on me
and as orange and hot
as sun at noonday practically
can be. Only this one
is straight at me like a

beam shot to hit me
It feels like
enforcing itself
on me giving me its
message
(*M* p. 577)

The moon stares *in* the window, the sun blazes *at* me. What is foregrounded here is less the determination of exterior objects by a subject than the experience of the world impinging upon the body by *including* it, as the poem includes this inclusion.

But within these inclusions, the particularity of the body, of the "me," also comes to feel itself sometimes *excluded*: abruptly distinguished by its mortality, cruelly specified by the finitude of its organic functioning, which qualifies its particular status as an object. In volume three of the *Maximus Poems*, being-for-death is not that which makes man world-forming while leaving material objects worldless, as Heidegger has it. Rather, it is one of the ways in which the formation of world impinges upon particular objects and inexorably distinguishes them, like a line break:

Ice
snow my car as hidden as a hut beneath it children passing without even notice, every house so likewise interesting because of snow upon each roof. Lamps, and day,
nothing not new and equally fresh forever upon this earth. All
but me, damned as each man in death itself the evil
which throws a dart of dirt and shadow on my soul and on
this Sunday when in this light, and on this point, no
conceivable hindrance would seem imaginable to darken
or in fact any difficulty of any sort except to keep
my eyes out of the sun-blaze on the sea and careful also
not to notice too directly the street, frozen and slippery as
the light
(*M* III.108)

Enjambment, in this passage, formally enacts the "hindrance" of abbreviated anticipation, the interruption of one entity among others by death, "a dart of dirt and shadow" in the otherwise new and fresh world of "All / but me." But if, for Heidegger, being-toward-death is what *essentially* separates "man" from "the stone," for Olson it is what at once distinguishes certain bodies while also rendering the life of those bodies relative to the objects of reality:

I am a stone

> or the ground beneath
>
> My life is buried,
> with all sorts of passages
> both on the sides and on the face turned down
> to the earth
> (*M* III.228)

Olson's most characteristic injunction is therefore:

> Love the World—and stay inside it.
> Concentrate
> one's own form, holding
> every automorphism(*M* III.188)

Olson's imperative is to *be in* the world while also concentrating "one's own form" and "holding / every automorphism," a double demand for both relation and distinction, integration and difference, which is answered by his notion of "objects in field."

This objectist imperative, shared by Whitehead and Olson, carries with it a corollary: that we recognize the being-open of any object as concomitant with its being-broken. To splice Olson's terms with Whitehead's, the community of actual things is an incompletion in process of production, and every automorphism therein depends upon the "proper confusions" of the societies by which it is composed. This is what Olson calls "the blessing / that difficulties are": no actual entity is ever equal to itself, insofar as its being itself depends upon its prehensions of and by other actual entities. Contra Pound, there is no ideal "room" in which a body does not cast a "shadow." *The world is bound by its brokenness and its boundedness is thus broken.*

The broken-boundedness of the world is a condition made eminently recognizable by volume three of the *Maximus Poems*, the unity of which is extremely fragile. In this sense, Olson's poetry is often truer to the contingent flexibility of his objectist "stance toward reality" than the tone of overbearing confidence that often characterizes his writing on poetics. The same is true of Pound's *Cantos*. The moment at which the *Cantos* "make sense" is precisely the moment at which Pound openly recognizes the "errors and wrecks" of his struggle to "make it cohere."[109] Likewise, the *Maximus Poems* achieve their most integral participation in the sense of the world when Olson's epic-making mythopoetic aspirations cave in, and the brokenness of the final volume manifests, in both form and content, the difficulties of "the way of / the lowest" (*M* 1.15).

Olson recognized the difficulties of objectism early in his career. "In Cold Hell, In Thicket" (1950) confronts the involvement of his objectist poetics with the problem of *objectification*:

> Who am I but by a fix, and another,
> a particle, and the congery of particles carefully picked one by another,
>
> as in this thicket, each
> smallest branch, plant, fern, root
> —roots life, on the surface, as nerves are laid open—
> must now (the bitterness of the taste of her) be
> isolated, observed, picked over, measured, raised
> as though a word, an accuracy were a pincer!
> this
> is the abstract, this
> is the cold doing, this
> is the almost impossible
>
> So shall you blame those
> who give it up, those who say
> it isn't worth the struggle?[110]

The struggle that Olson specifies is not to save the organism from the condition of the object but rather to save the object from objectification. The "abstract" or the "cold doing" to which the poem refers is the isolation of any smallest unit from the relational collectives in which it is included, and the concomitant failure to recognize that any supposedly "simple location" is always also elsewhere, through its processual involvement with other locations and occasions. We effectively "give up" the struggle that the poem specifies if we try to overcome the "proper confusions" of objects by making organizational closure or organic holism our criterion of coherence.

The broken-boundedness of the common world thus entails the difficulty of sustaining a struggle between the wavering of what is, the precisions of pincers, and the accuracy of every automorphism, without capitulating to a condition of closure:

> as even the snow-flakes waver in the light's eye
>
> as even forever wavers (gutters
> in the wind of loss)

even as he will forever waver

precise as hell is, precise
as any words, or wagon,
can be made[111]

. . .

And the too strong grasping of it,
when it is pressed together and condensed,
loses it

this very thing you are[112]

If "a word" or "an accuracy" is not to operate as "a pincer," then the flexibility of a field of tensions will have to be sustained, by any practice of making, against "the too strong grasping of it," which loses "this very thing you are." The desideratum of the poem seems to be a practice of fabrication involving *tentative* precisions; it recognizes that our relation to other objects is constitutive of our own integrity and that the latter is dissolved as soon as the wavering of those relations is elided. If I have emphasized the brokenness of the *Maximus Poems*, the becoming-fragile of their final pages as constitutive of the open field they form, that is because what may matter most at the limits of fabrication in either technoscience or poetry is not to forget the *fallibility* of matter—a negative capability, a persisting in doubts and uncertainties, that is inherent to the "congery of particles" of which the world is composed.

Perhaps we have become so accustomed to thinking of Olson as a gigantic bombastic mythologizer that we often forget the profound tenderness that is pervasively evident in the content of his poems, in his handling of rhyme, in his sense of the line, in the sense of the world that his writing makes manifest. So that we don't forget, we should read the entirety of "West Gloucester," which is printed on the page of the *Maximus Poems* immediately following Olson's invocation of "the outer limits" of "the electro-magnetic strain":

Condylura
cristata
on Atlantic
Street West
Gloucester
spinning on its
star-wheel

nose,
in the middle of the
tarvia,

probably because it had been
knocked in the head (was
actually fighting all the time,
with its fore-paws at
the lovely mushroom growth
of its nose, snow-ball flake pink flesh
of a gentian, until I
took an oar out of the back seat of the station wagon
and removed it
like a pea on a knife to
the side of the road

stopped its dance dizzy dance
on its own nose out of its head
working as though it would get rid of
its own pink appendage

like a flower dizzy
with its own self

like the prettiest thing in the world drilling
itself into the

pavement

and I gave it, I hope, all the marshes of Walker's Creek
to get it off what might also seem
what was wrong with it, that the highway
had magnetized the poor thing

the loveliest animal I believe I ever did see
In such a quandary

and off the marshes
of Walker's Creek fall

graduatedly so softly to
the Creek and the Creek to
Ipswich Bay an arm
of the Atlantic Ocean

send the Star Nosed Mole
all into the grass
all away from the dizzying
highway if that was what was wrong

with the little thing, spinning
in the middle of the
highway

(*M* III.26–27)

Wavering from the center of the page to the left margin, "like a pea on a knife to / the side of the road," and then back to "the middle of the / highway," the poem likewise moves between the discourse of scientific taxonomy ("Condylura / cristata) and poetic metaphor ("snow-ball flake pink flesh / of a gentian"). It draws together regional and global geography—"Walker's Creek" and "the Atlantic Ocean"—through the central figure of a creature at once cosmological and mundane, local and exotic: the Star Nosed Mole. The poem concerns an encounter, and the speaker keeps returning to the moment of this encounter as if to punctuate the poem, while also speculating upon its cause and narrating its consequences. The poem includes a hypothesis ("probably because it had been / knocked in the head"), an intervention ("until I / took an oar out of the back seat of the station wagon / and removed it"), a tribute ("like a flower dizzy / with its own self // like the prettiest thing in the world"), and a benediction ("send the Star Nosed Mole / all into the grass / all away from the dizzying / highway"). Crucially, the poem involves a relation between the *disorientation* of the mole and the *tentative precision* characterizing the speaker's intervention. It holds this disorientation, this tentativeness, and this precision together, as a complex or a problem, in the singularly awkward beauty of a couplet whose rhyme transpires as though also contingently stumbled upon, as though itself an encounter: "the loveliest animal I believe I ever did see / in such a quandary."

The world does not depend upon us, but we have no choice but to intervene in the world, in "nature," because like any other object we *are* an intervention. The question that concerns Olson is: *How* do we intervene in the congery of particles that the world is, that we are, without making pincers of our precisions, a cold hell out of our accuracies? As they manifest themselves in "West Gloucester," the tentative imperatives of Olson's poetry seem to be as follows: our localism cannot be parochial in its gradations of relevance, because there is no *certain* scale to which our actions are appropriate. We have to specify that which we encounter, whether Condylura cristata, Atlantic Street West, Walker's Creek, Ipswich Bay, the Atlantic Ocean, snow-ball flake pink flesh, or the minutiae that hold and swim at the outer limits of the electromagnetic strain. We have to intervene in our encounters with care, and with whatever irreparably perfect awkward instruments present themselves so delicately to the occasion. And we have to be provisional. As Derrida reminds us somewhere, we must always say "perhaps" (*peut-être*). And as Olson so carefully suggests, it is not necessarily contradictory to our accuracies or our actions to say "probably because," "I hope," "what might also seem," "if that was what was wrong."

3

DESIGN SCIENCE: GEODESIC ARCHITECTURE IN NANOSCALE CARBON CHEMISTRY AND RONALD JOHNSON'S *ARK*

The most enduringly reproducible design entities of Universe are those occurring at the min-max limits of simplicity and symmetry.

—Buckminster Fuller, *Synergetics*

Perhaps no other building combines the fundamentals of structural design and architecture more organically than does the geodesic dome of R. Buckminster Fuller.

—Harold Kroto, "Macro-, Micro-, and Nano-scale Engineering"

look up Bucky Fuller's
synergetics, etc.

—Ronald Johnson, Notebooks

Nineteen ninety-one was marked by two signal events in materials science and materialist poetics. At NEC Laboratories, Japanese physicist Sumio Iijima discovered "carbon nanotubes," rolled sheets of hexagonal carbon molecules that, artificially synthesized, are now among the most important and commonly used nanoscale materials. In the same year, Ronald Johnson completed his long "architectural" poem, *ARK*, on which he had been at work since the early 1970s. What these events have in common is a history linking their development, their *Bildung*, to that of Buckminster Fuller's geodesic architecture. Excavating this common ground, I want to trace the influence of Fuller's "design science" upon the structural articulation of these materials—carbon nanotubes and concrete poetry—as one instance of the way in which materials science and materialist poetics, as practices of fabrication, are traversed by common models of material construction. At the center of this story will be the concept (the ideology, in fact) of "design" and its relation to a certain idealist concept of "nature" and the "nature poem." What is at issue is the tension between *patterns* and *materials* agitating the desiderata of formal efficiency and structural perfection in techno-science and poetry.

Johnson describes *ARK*, a major long poem on which he worked for twenty years, as "literally an architecture . . . fitted together with shards of language, in a kind of cement music."[1] The complete volume is comprised of three books, *The Foundations*, *The Spires*, and *The Ramparts*, each consisting of thirty-three "beams," "spires," and "arcades" (respectively). Perhaps the primary literary historical significance of Johnson's poem is that it emerges equally from the immediate influence of Charles Olson and Louis Zukofsky, thus confounding any stark opposition between "open field" and "procedural" traditions in postwar poetry and poetics. *ARK* combines prose meditations on physics and biology, procedural rewritings of a multiplicity of texts, ecstatic paeans to the prolific complexity of the natural world, and impeccably crafted "verbivocovisual" concrete poetry. As Marjorie Perloff notes, this last element of Johnson's work "marks a key transition between the more programmatic visual Concretism of the 1950s and 1960s and the visual/sound poetries of the last two decades, especially those of Susan Howe and Johanna Drucker, Christian Bök and Brian Kim Stefans."[2] Spanning the period of its production from the 1970s to the 1990s, *ARK* serves as a literary historical bridge between the mid-twentieth-century context from which it stems (Concretism, Olson, Zukofsky) to the visual/sound poetry of Christian Bök and Caroline Bergvall to which I will turn in Chapter 4.

The "carbon nanotubes" discovered by Sumio Iijima in 1991 are the strongest materials that we know of and, in terms of engineering applications, among the most promising nanoscale materials yet produced. They are "a material some fifty to one hundred times stronger than steel at about one sixth the weight,"[3] with applications running from use in cosmetics, textiles, and polymer composites to functioning as low-resistance wires on molecular computer chips. They are likely candidates to serve as field emitters in wearable media applications, and they already function as components of prototype composite materials and nanoscale sensor technologies. Nanotechnologists have not been shy in promoting the significance of nanotubes for the future of materials research and fabrication: Nobel laureate Harold Kroto argues that their extraordinary structural efficiency and their multifunctional physical properties "could revolutionize civil and electronic engineering as much as steel and aluminum did in earlier times."[4]

Kroto offers these remarks about carbon nanotubes—or "buckytubes," as he calls them—in his contribution to the volume *Buckminster Fuller: Anthology for the New Millennium*. "These amazing new materials," he notes, "hark back to the patents of Buckminster Fuller."[5] In 1984, midway through the construction of his architectural poem, Ronald Johnson declares that an earlier work, *RADI OS* (his "ventilated" version of the opening four books of *Paradise Lost*), will serve as the one-hundredth and final section of *ARK* upon the latter's completion—as "a kind of Dymaxion Dome over the whole."[6] But to understand how the development of carbon nanotubes and the composition of *ARK* draw common inspiration from the "Dymaxion" principles of Buckminster Fuller's "design science," we need to step back from 1991 to 1948: to the summer at Black Mountain College just prior to Charles Olson's first term as an occasional instructor.

BUCKMINSTER FULLER

It is a staple of two interlocking mythologies, that of a paradise of radical pedagogy and interdisciplinary collaboration just prior to the odious dominion of one-dimensional corporate America in the 1950s, and that of a self-made renaissance man poised at the crux of his transition from misunderstood genius to international icon: the story of how Buckminster Fuller's first geodesic dome was (momentarily) assembled at Black Mountain College in the summer of 1948.[7] Arriving at the college for his first term as an invited instructor, Fuller brought along a mobile home trailer packed with glass spheres and geometrical models. He ran an architecture/design workshop through the 1948 Summer Session in the Arts, spinning out three-hour lectures on energetic geometry to students and a faculty that included Josef and Anni Albers, Elaine and Willem de Kooning, Ruth Asawa, John Cage, and Merce Cunningham. Elaine de Kooning's account of the first of those lectures testifies to the impact of Fuller's design science upon the college of which Olson would become rector in 1951:

> Bucky whirled off his talk, using bobby pins, clothespins, all sorts of units from the five-and-ten-cent store to make geometric, mobile constructions, collapsing an ingeniously fashioned icosahedron by twisting it and doubling and tripling the modules down to a tetrahedron; talking about the obsolescence of the square, the cube, the numbers two and ten (throwing in a short history of ciphering and why it was punishable by death in the Dark Ages); extolling the numbers nine and three, the circle, the triangle, the tetrahedron, and the sphere; dazzling us with his complex theories of

> ecology, engineering, and technology. Then he began making diagrams on a blackboard. He drew a square, connecting two corners with a diagonal line. "Ah", he said affectionately, "here's our old friend, the hypotenuse."[8]

"The whole world," John Cage would say, "has to be turned into music or into a Fuller university" (Fig. 12).[9]

Having transformed Black Mountain into such a university, Fuller spent the summer developing an alternative to the obsolescent cube: a model dome, four feet in diameter and based on the division of a sphere into triangularly interlocking circumferences. At the end of the Summer Session, this model was to serve as the basis of a full-size structure spanning forty-eight feet, to be built out of narrow, aluminum alloy Venetian-blind slats. These materials were all Fuller or the college were able to afford at the time. With his fellow instructors looking on, Fuller and his students scrambled around a field in the rain, laying out the strips according to the pattern of the model and fastening them at their vertices (Fig. 13).

There are varying reports of the outcome of this project. According to the narrative offered by Robert Marks in *The Dymaxion World of Buckminster Fuller*,

> he intentionally designed this structure so that its delicate system gently collapsed as it neared completion. He then fortified the individual chords of the triangular system with prismatically arranged additions of two more Venetian blind strips. Gradually the structure reassumed its domical configuration. . . . Fuller here arranged to bring the structure up to critical capability by the gradual addition of discrete increments. The result was that the safe structure of the 48-foot dome was accomplished with one-hundredth of the weight of material customarily employed.[10]

Most versions of this myth of origin, however, do not include the resurrection of a "domical configuration" after the structure's initial collapse. Merce Cunningham's account of Fuller's demonstration accords it a rather more modest accomplishment than the architectural triumph Marks describes:

> Yes, he built one of those geodesic domes, and my recollection about it is that he asked for a specific kind of material but the width was less. So he was heard very quietly to make this remark as they were trying to put it up: "Well it won't work but we'll go right ahead and do it anyway." And it did fall down. But it was such an experience for any of us who watched.[11]

Here Fuller succeeds in offering delight and instruction, but not quite in fabricating an architectural system of unprecedented structural efficiency. Despite

Figure 12 Buckminster Fuller at Black Mountain College, 1949. The first prototype of a geodesic dome hangs toward the top left of the frame. Courtesy of The Estate of Buckminster R. Fuller.

these inconsistencies of historical record, however, what is emphasized by all accounts of Fuller's attempted construction of his first geodesic dome in the summer of 1948 is an effort to create a stable structure based on the complex modeling of abstract geometrical principles with an absolutely minimal volume of material. Both the attempt itself and the discrepancies in its reportage are important because they exemplify a core conceptual tension underlying all of Fuller's thinking, a tension particularly evident in the relation between his concepts of "synergy" and "ephemeralization."

Figure 13 Buckminster Fuller with students and instructors attempting to raise the first geodesic dome. Black Mountain College, 1948. The four-foot prototype of the dome is shown on the left. Courtesy of The Estate of Buckminster R. Fuller.

"Synergy," Fuller explains at the outset of *Synergetics*, his 1,500-page, two-volume set of *Explorations in the Geometry of Thinking*, "means behavior of whole systems unpredicted by the behavior of their parts taken separately. Synergy means behavior of integral, aggregate, whole systems unpredicted by behaviors of any of their components or subassemblies of their components taken separately from the whole."[12] "Ephemeralization," Fuller notes in *Critical Path*, his handbook for "making humans a success in Universe," involves "the invisible chemical, metallurgical, and electronic production of ever-more-efficient and satisfyingly effective performance with the investment of ever-less weight and volume of materials per unit function formed or performed."[13] While the principle of ephemeralization would seem to privilege the production of lighter, stronger, and more functional *materials* as the key to industrial progress, the principle of synergy promotes the primacy of a *system* of behavior that is irreducible to any particular material instantiation.

These two principles converge, however, in the example of "chrome-nickel-steel" offered by Fuller in *Synergetics*. "Synergy alone explains metals increasing in their strength," he writes. "All alloys are synergetic. Chrome-nickel-steel has an extraordinary total behavior. In fact, it is the high cohesive strength and structural stability of chrome-nickel-steel at enormous temperatures that has made possible the jet engine." This synergetic coherence, Fuller stipulates, disproves the principle that "a chain is only as strong as its weakest link." Whereas the combined tensile strengths of the elements compounded in chrome-nickel-steel amounts to only 260,000 psi, the tensile strength of the alloy itself is a minimum 350,000 psi. "Here we have the behavior of the whole completely unpredicted by the behavior of the parts," Fuller concludes. "The augmented coherence of the chrome-nickel-steel alloy is accounted for only by the whole complex of omnidirectional, intermass-attractions of the crowded together atoms."[14] It is the *complex* formed by the association of atoms, the pattern of their relations and not the atoms themselves, that enables the unexpected combination of lightness, strength, and structural integrity that makes alloys exemplary instances of ephemeralization.

The case of alloys should make it clear that materials and whole systems are reciprocally determinate. A consideration of the distinct components chrome, nickel, and steel is inadequate to predict the synergetic behavior in which their combination results, but it is nonetheless only by combining *these* components that such synergetic behavior is produced. Yet in reflecting upon Fuller's admirably dogged attempt to raise the first geodesic dome at Black Mountain College, despite the knowledge that his materials were simply inadequate to the load-bearing requirements of the structure, one might detect at least the trace of a conviction that if the geometry is sufficiently harmonious then structural integrity will take care of itself. In other words, we might detect the shadow of an intimation that materials make no difference at all.

Indeed, Fuller's rhetoric tends decidedly in this direction. "There are no solids," he writes. "There are no things. There are only interfering and noninterfering patterns operative in pure principle, and principles are eternal. Principles never contradict principles."[15] Note the irreconcilability of this perspective with Olson and Creeley's objectism, and also with the primacy of "actual entities" in Whitehead's cosmology. Whitehead and Fuller would both agree that all things are "complex and interdependent."[16] But for Fuller, this complex interdependency ("patterns operative in pure principle") apparently displaces the things themselves. For Whitehead, on the other hand, complex interdependencies are what *constitute* actual entities, which "are the final real things of which the world is made up." Thus he insists that "there is no going behind actual entities to find anything more real."[17] If Olson, cognizant of the post-classical physics upon which Fuller's rejection of solids relies, places the word "solid" in scare quotes in "Projective Verse," he nonetheless insists that an open field is a field *of objects* and that "it is a matter, finally, of OBJECTS" rather than of patterns.[18] It is worth returning as well to Creeley's elaboration of his formula, "Form IS the extension of content":

> The use, USE, by any means whatever, anything that will work, of MATERIAL (sky man horse, steel iron coal, potato onion carrot, H20 analysis fission) MATERIAL (OBJECT), to make known, to DECLARE: CONTENT. Fused, driven thru, beaten, hammered, fired: FUSED—IS FORM, is the FORM. No "patterns", nothing but what is SO MADE.[19]

"No 'patterns,'" Creeley insists. For Creeley—as for Fuller, to be sure—any "material" is a fused coagulation of components. But for Creeley the "form" of this coagulation emerges precisely from a process of contradiction (it is "beaten, hammered, fired"), not from patterns operative in pure principle, nor from eternal principles that never contradict.

The conflict, indeed the contradiction, between objectism and synergetics thus recapitulates the contradiction between objectism (Olson, Creeley) and organic form (Levertov, Duncan) that we have already articulated. Both of these contradictions are *internal* to the milieu of Black Mountain College. If we were to look for a synthetic term, drawing upon both of these trajectories at Black Mountain, it might be something like the poetics of another Black Mountain poet, but one who was certainly closer to Fuller than to either Olson or Creeley: John Cage. In Cage's work we have systems of aleatory procedures that disrupt pattern formation while nonetheless situating "the objects of the poem" within a formal *system*, rather than what Olson would call a field. Cage's practice of composition involves schemata of possible chance operations serving as a closed system within which indeterminacies play out. This is not a criticism of his practice, since a predetermined system of constraints and of chance operations can be generative of novelty. In all cases—those of Fuller, Olson, or Cage—we are no doubt dealing with "a series of tensions" that "are made to hold." At issue is the operative basis for the cohesion of a "series": that of *patterns* operative in pure principle (Fuller), a determinate *system* within which indeterminate relations unfold (Cage), or a flexible *field* constituted by the permeable relationality of objects (Olson). As we shall see, it is somewhere in the interstices of these distinctions that Ronald Johnson will compose the series of tensions called *ARK*.

The primacy of "patterns operative in pure principle" in Fuller's thought prepares us for his concept of *design*. Since it is a choice specimen of the fraught prose style through which Fuller filters his ideas of order, it is worth quoting in full the definition of "design" offered in *Synergetics*:

> The discovery by human mind, i.e., intellect, of eternally generalized principles that are only intellectually comprehendable and only intuitively apprehended—and only intellectually comprehended principles being further discovered to be interaccommodative—altogether discloses what can only be complexedly defined as a *design*, design being a complex of interaccommodation and of orderly interaccommodation whose omni-integrity of interaccommodation order can only be itself described as intellectually immaculate. Human mind (intellect) has experimentally demonstrated at least limited access to the eternal design intellectually governing eternally regenerative Universe.[20]

From this passage, we can extract the following traits of "design" as Fuller understands it. First, design is *eternal*: it involves "eternally generalized principles." Second, design is *universal*: its principles are general, and only those principles that are found to be "interaccommodative"—which are compossible, or noncontradictory—count as eternally generalized. Third, these principles are *intellectually accessible*: they are accessible to comprehension (intellect) and apprehension (intuition). Fourth, following from the compossibility of its principles, design is fully *integrated*: it is a "complex of interaccommodation." Fifth, this complex of interaccommodation is *orderly*: it is gathered into a total pattern, or an "omni-integrity of interaccommodation order." Sixth, this order is *perfect*: it is "intellectually immaculate," suggesting its production by a higher intellect than "human mind," since the latter has thus far exhibited only "limited access to the eternal design." Seventh, design *governs* the totality of that which is: there is an eternal design "intellectually governing eternally regenerative Universe." To summarize, design is an eternal, universal, and perfectly ordered integrity of intellectually accessible principles governing all that is.

It is such an understanding of "design" that undergirds Fuller's quest, in the two years of intensive research prior to his summer at Black Mountain in 1948, to arrive at an understanding of "the comprehensive coordinate system employed by nature"[21] and to formulate "a finite system in universal geometry."[22] If this goal sharply distinguishes his sensibility from Olson's hatred of universality, Fuller's exposition of his motives in this pursuit nonetheless attests to their common aversion to academic "disciplines," an aversion linking both Fuller and Olson to the polymathic ethos of Black Mountain College:

> About 1917, I decided that nature did not have separate, independently operating departments of physics, chemistry, biology, mathematics, ethics, etc. Nature did not call a department heads' meeting when I threw a green apple into the pond, with the department heads having to make a decision about how to handle this biological encounter with chemistry's water and the unauthorized use of the physics department's waves. I decided that it didn't require a Ph.D. to discern that nature probably had only one department and only one coordinate, omnirational, mensuration system.[23]

But the distinction we should make between Fuller and Olson, on this point, is that for Olson the problem is not that the universe is too fragmentary but that it is not fragmentary enough. Olson abhors the idea of any univocal system of measure. He does not propose that there is *one* omnirational mensuration system but, on the contrary, that *every single* actual entity is a field of measure. We might call this *metrontology*: an understanding of *determinate being* as measure. This is the import of a question posed by Olson to Fuller in the summer of 1949, which he reports in a letter to Robert Creeley written from Black Mountain College in 1952: "I drove Buckminster Fuller out of my house here three years ago come summer by saying to that filthiest of all the modern design filthers: 'In what sense does any extrapolation of me beyond my fingernails add a fucking thing to me as a man?'"[24] Olson's distaste for Fuller's omnirational mensuration system is commensurate with his hatred of "design," the "eternally generalized principles" of which fail to recognize the particularity of measure.

If Olson's investigations of composition by field in early poems like "La Preface," "The Praises," and "The Kingfishers," along with his correspondence with Creeley in the late 1940s, led to the making of the *Maximus Poems*, it was Fuller's research into energetic geometry from 1946 to 1949 that led most directly to his patent for the geodesic domes that would make him famous in the 1950s.[25] Experimenting with the closest packing of spheres, he noted the elementary fact that concentric arrangements of spheres around a center give rise to hexagonal arrays with triangular subdivisions if the spheres are joined at their centers by straight lines. He explored the implications of the fact that the packing of spheres on a plane in successive layers around a nucleus generates a series of six, then twelve, then forty-two, then ninety-two components, and that this finite pattern of symmetrical concentric organization then begins to repeat itself infinitely if the procedure is continued "omnidirectionally." (This satisfies Fuller's definition of a *system*: "a patterning of force that returns upon itself in all directions—that is, a closed configuration of vectors.")[26] And he generated what he called the "Vector Equilibrium"—a fourteen-sided polygon with eight triangular and six square faces—by stacking spheres into a configuration of three vertical layers consisting of three, six, and three components around a central sphere and then joining all their central points in a system of vertices (Fig. 14).

But it was Fuller's work during this same period on the division of a sphere into "great circles" or "geodesics"—equators running the entire circumference of the sphere—that resulted in his first dome prototype. By modeling the rotation of an icosahedron around the interior of a sphere, he arrived at thirty-one great circles crisscrossing the sphere symmetrically and dividing it entirely into a network of triangles. Cutting such a sphere in half gave him the model for the structure that he attempted to raise in 1948 (Fig. 15).

In the summer of 1949 (during which Olson apparently drove him out of his house) Fuller returned with a new foldable dome structure, this time only fourteen feet in diameter, that combined elements of his research into great circles and the closest packing of spheres. In this case the same network of interlocking triangles was composed of an articulated system of straight tubes with interior cables, rather than continuous curved strips, held together at nodal points by circular joints. And rather than Venetian-blind slats, the structure was built out of high-quality aluminum aircraft tubing (Fig. 16). Here is the description of this "necklace dome" offered by Hugh Kenner in his 1973 study of Fuller's work, *Bucky*:

> [In 1949] Bucky was threading cables through tubes, to make an intricate necklace that lay on the ground until the cables were tightened. But when they were tensioned to draw the tubes together, the structure erected itself into a dome shape. A photo shows nine men hanging from it. Their weight tended to compress the tubes, which ran vertex to vertex. Their weight also stressed the tension network. Since slacking the cables would let the whole thing collapse, and omitting the tubes would leave only a structureless net, the demonstration of tensional and compressional interplay could not have been neater.[27]

It was also during Fuller's second summer at Black Mountain that his student, Kenneth Snelson, modeled the "tensegrity principle" (Fuller's term: tension + integrity), a structure that was already nascent in the network of tubes and cables operative in the necklace dome. But while Fuller's dome compressed *through* the intersection of the tubes, models based on Snelson's designs distributed tensional and compressive forces such that none of the component tubes strung along a network of wires touch one another at all (Fig. 17).

For Fuller, the discovery of such structures exemplifies the human mind's intellectual access to eternal design. "Nothing in Universe touches anything else," he writes:

> The Greeks misassumed that there was something called a solid. Democritus thought it could be that there were some smallest things in those solids, to which he gave the name atom. Today we know that the electron is as remote from its nucleus as is the Earth from the Moon in respect to their diameters. We know that microcosmically and macrocosmically nothing touches.[28]

The vast geodesic domes that Fuller would build in the 1950s and 1960s—combining his research into great circles, the close-packing of spheres, and tensegrity—would embody this principle. Or at least, they would embody it *in principle.*

"Tensegrity structures are the essence of all geodesic domes," Fuller writes. But whenever an essence is embodied in a *material* structure, one will eventually have to account for some deviation from first principles:

> Because the materials used in the construction of the dome have some substantial dimension, we get to the point where the high-frequency production of the arc-altitude is such that the materials (the individual tensegrity components) touch one another. Every one of these elements is where it wants to be within the structure—there are no tensions anywhere, no slacks, all of the stresses are absolutely even—so we then fasten the two structural components together where they touch and where they want to be. This takes out the springiness of the geodesic domes and makes them rigidified. Because the tension and tensegrity have no limit of clear spanning, tensegrity structures open up completely clear-spanned dome structures of any size.[29]

Such clear-spanned structures are the ideal resolution of the primary "tension" in Fuller's thought, manifested by the collapse of his first geodesic structure at Black Mountain. Because they "have some substantial dimension," *the materials* put some pressure on the *essence*—"the omni-integrity of interaccommodation order"—that a structure is designed to embody. Although, in principle, nothing touches anything else, Fuller acknowledges that the materials nevertheless *do* touch one another. But the pressure they exert upon each other, and upon principles, is balanced by a network of tensions. Supposedly, this network of tensions is itself so perfectly balanced that it alleviates any tension at

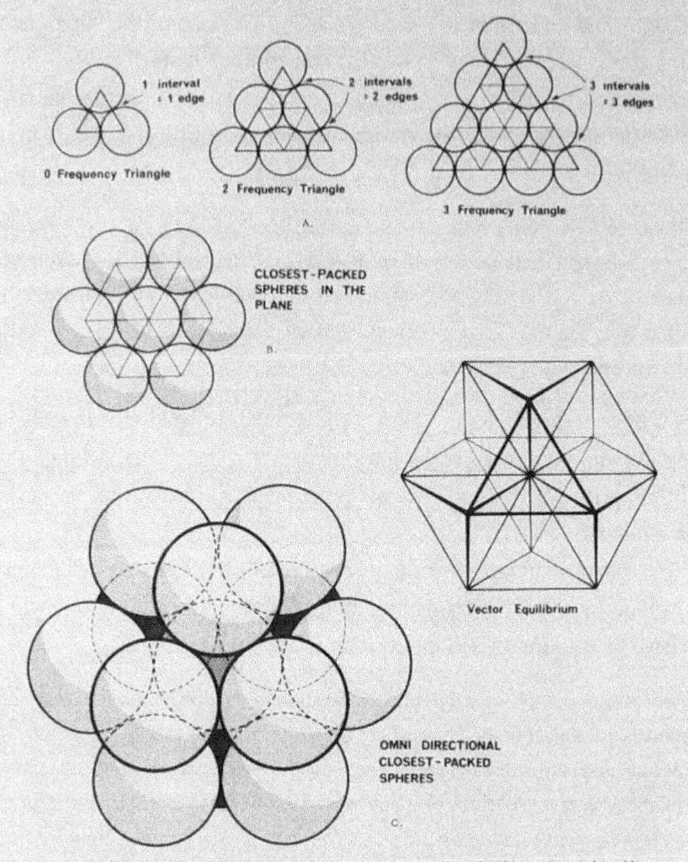

Fig. 413.01 *Vector Equilibrium: Omnidirectional Closest Packing Around a Nucleus:* Triangles can be subdivided into greater and greater numbers of similar units. The number of modular subdivisions along any edge can be referred to as the *frequency* of a given triangle. In triangular grids each vertex may be expanded to become a circle or sphere showing the inherent relationship between closest packed spheres and triangulation. The frequency of triangular arrays of spheres in the plane is determined by counting the number of intervals (A) rather than the number of spheres on a given edge. In the case of concentric packings or spheres around a nucleus the frequency of a given system can either be the edge subdivision or the number of concentric shells or layers. Concentric packings in the plane give rise to hexagonal arrays (B) and omnidirectional closest packing of equal spheres around a nucleus (C) gives rise to the vector equilibrium (D).

Figure 14 Generation of the vector equilibrium from closest packing of spheres. From Buckminster Fuller, *Synergetics.* Courtesy of The Estate of Buckminster R. Fuller.

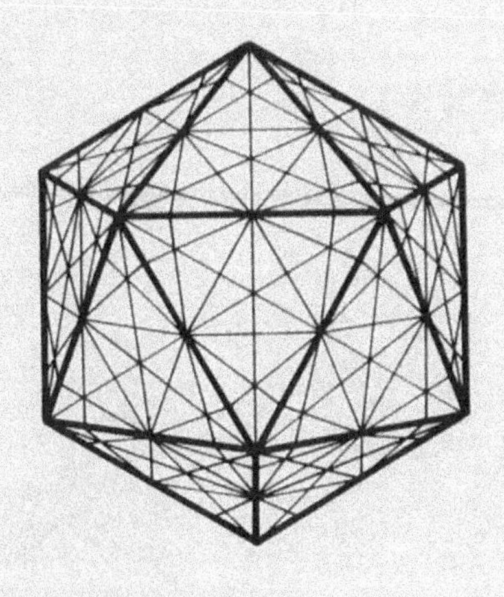

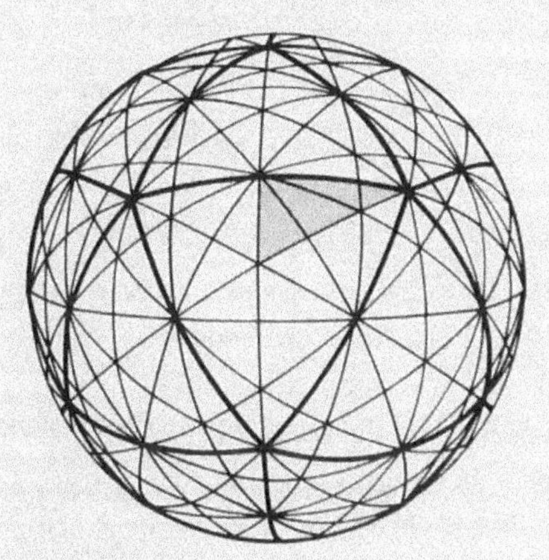

Fig. 457.30B *Projection of 31 Great-Circle Planes in Icosahedron System:* The complete icosahedron system of 31 great-circle planes shown with the planar icosahedron as well as true circles on a sphere (6 + 10 + 15 = 31). The heavy lines show the edges of the original 20-faced icosahedron.

Figure 15 Projection of 31 great-circle planes in an icosahedron system. From Buckminster Fuller, *Synergetics*. Courtesy of The Estate of Buckminster R. Fuller.

Figure 16 Buckminster Fuller and students hanging from "necklace dome." Black Mountain College, 1949. Courtesy of The Estate of Buckminster R. Fuller.

all. In a geodesic dome, "there are no tensions anywhere, no slacks, all of the stresses are absolutely even." The domes are icons of a perfect equilibrium, and they instantiate the *teleological* vector of design.

We can thus measure, vis-à-vis the relation of materials to the teleology of equilibrium, the distance between two approaches to fabrication emerging from Black Mountain: design science and objectism. In 1950, one year after Fuller's last summer at the college and one year prior to Fuller's patent for geodesic domes, Olson would affirm the "proper confusions" of objects in "Projective Verse," and he would state that "the objects which occur at every given moment of composition . . . are, can be, must be treated exactly as they *do* occur therein."[30] On the other hand, Fuller tells us that design is achieved when the materials are just exactly where they *want* to be.

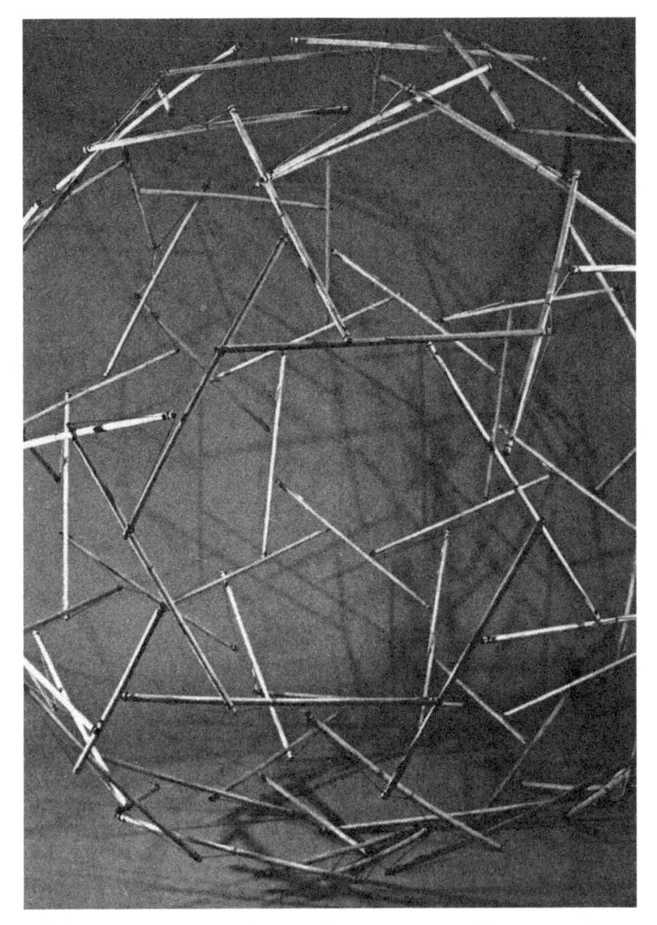

Figure 17 Tensegrity hemisphere, ca. 1959. Courtesy of The Estate of Buckminster R. Fuller.

BUCKMINSTERFULLERENE

Sometime in 1861 or 1862—so goes another famous myth of origin in the history of design—August Kekulé, an aspiring architect turned chemist, dreamed of a serpent swallowing its tail. He awoke to the realization that the properties of the benzene molecule resulted from a cyclical organization: a hexagonal arrangement of six carbon atoms with their six associated hydrogen atoms distributed around the vertices (Fig. 18).

In *Gravity's Rainbow*, Thomas Pynchon narrates this myth of origin as follows:

> Young ex-architect Kekulé went looking among the molecules of the time for the hidden shapes he knew were there, shapes he did not like to think of as real physical structures, but as "rational formulas," showing the relationships that went on in "metamorphosis," his quaint 19th-century way of saying "chemical reactions." But he could visualize. He *saw* the four bonds of carbon, lying in a tetrahedron—he *showed* how carbon atoms could link up, one to another, into a long chain. . . . But he was stumped when he got to benzene. He knew there were six carbon atoms with a hydrogen attached to each one—but he could not see the shape. Not until the dream: until he was made to see it, so that others might be seduced by its physical beauty, and begin to think of it as a blueprint, a basis for new compounds, new arrangements, so that there would be a field of aromatic chemistry to ally itself with secular power, and find new methods of synthesis.[31]

Whether one approaches Kekulé's retrospective reportage of his dream with a sense of paranoid foreboding or with a more detached mode of suspicion,[32] few would dispute that his 1865 article on the structure of the benzene ring marked a crucial turning point in the transition from analytic to synthetic chemistry.[33] Chemists could now systematically *fabricate* a vast family of derivative molecules based on a precise understanding of the spatial arrangement of atoms—of their topological distribution—and of the substitutions their arrangement enabled.

One hundred and twenty years later, in 1985, chemists Harold Kroto, Richard Smalley, and Robert Curl were trying to account for the inordinate regularity with which they encountered C60, a cluster of exactly sixty carbon atoms, among the molecular products of experiments involving the vaporization of graphite. Postulating that the high frequency with which this particular cluster showed up in their results was attributable to the superior stability of a highly symmetrical organization, they set about attempting to discern the possible structure of such a molecule. Kroto initially proposed a layering of graphite-like sheets of carbon, hexagonally organized and symmetrically stacked. But such a structure leaves unbonded electrons at the edge of the sheets, "dangling bonds," which would make the cluster highly reactive and hence unstable.[34]

It was while thinking through the possibility that such sheets of hexagonally organized carbon might spontaneously curl in upon themselves rather than falling into layers that Kroto (who was once an aspiring designer)[35] recalled his visit to what remains the largest extant geodesic structure: Buckminster Fuller's

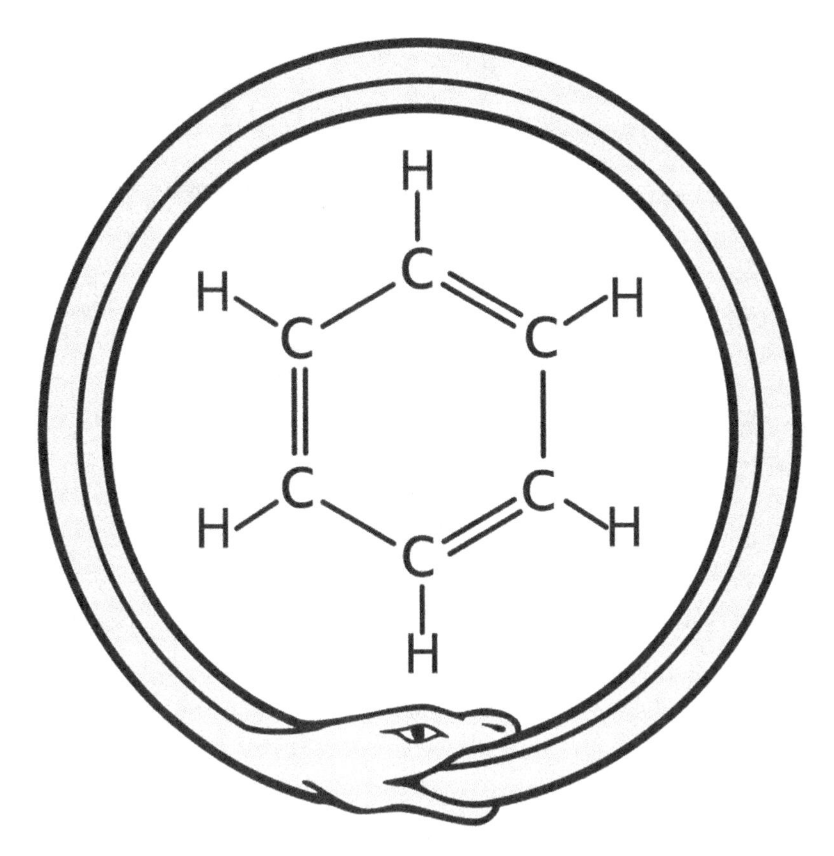

Figure 18 Ouroboros encircling the atomic structure of the benzene molecule. Six carbon atoms are bonded in a hexagonal arrangement with six hydrogen atoms at the vertices (double lines represent double bonds). Courtesy of D. M. Gualtieri under Creative Commons ShareAlike 3.0.

Expo Dome, constructed for the 1967 Montreal World's Fair. If the decisive clue to the riddle of the benzene molecule was supposedly Kekulé's dream of the Ouroboros, then, so the story goes, the clue to the riddle of C60 was Kroto's "vision of the dome" (Fig. 19).[36]

But the problem with curling hexagonally organized components into spherical cages is that it doesn't work: eventually the facets of the hexagons will fall out of alignment before they close. As D'Arcy Thompson noted in his classic *On Growth and Form* (1927), and as Buckminster Fuller knew, Leonard Euler had demonstrated in 1752 that

Figure 19 Buckminster Fuller's Expo Dome, constructed for the American Pavilion of the 1967 World's Fair in Montreal. Courtesy of Eberhard von Nellenburg under GNU Free Documentation License.

> *no system of hexagons can enclose space*; whether the hexagons be equal or unequal, regular or irregular, it is still under all circumstances mathematically impossible. So we learn from Euler: the array of hexagons may be extended as far as you please, and over a surface either plane or curved, but *it never closes in*.[37]

The predominantly hexagonal structures of the radiolaria studied by Thompson also included *pentagons*, as did Fuller's Expo Dome (Fig. 20). Kroto mentioned something about this detail to his colleagues, along with the fact that pentagons were involved in the small "stardome," designed by Fuller, that Kroto had constructed for his children years before at his home in England.[38] But, not immediately recognizing that these pentagonal components were the key to the geometrical problem facing the research team, he did not emphasize this point (Fig. 21). Nonetheless, Kroto's suggestion that the cluster might be structurally homologous to Fuller's domes led Richard Smalley to withdraw *The Dymaxion World of Buckminster Fuller* from the Rice University library. Had he attended carefully to an image of Fuller surrounded by his students at Black Mountain

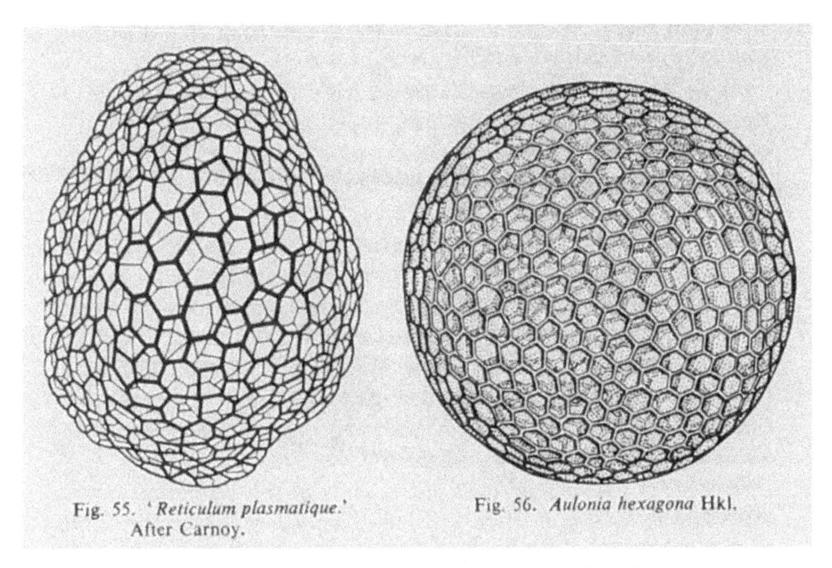

Figure 20 Drawings of radiolaria by Ernst Haeckel included in D'Arcy Thompson's *On Growth and Form.* Can you spot the pentagonal components of the structures?

College in the summer of 1948, pointing directly to a pentagonal face among the hexagons of the first prototype for a geodesic dome, Smalley would immediately have found the answer he was looking for (Fig. 22). Unable, however, to spot the pentagonal components of the other domes upon which he was apparently focused, Smalley returned home and began constructing molecular models out of paper hexagons, all of which were failures, until he eventually remembered Kroto's stray reference to the pentagons in Fuller's stardome. Prompted to cut out several five-sided polygons and add those to his model, it wasn't long until Smalley had discovered a closed configuration with exactly sixty vertices.[39]

On November 14, 1985, the research team published a letter in *Nature* announcing the discovery of a perfectly symmetrical molecule, one nanometer in diameter, consisting of sixty carbon atoms arranged in a polyhedral structure of twelve pentagons and twenty hexagons. That is precisely the structure of a soccer ball, but the title the authors gave to their now famous paper was "C60: Buckminsterfullerene" (Fig. 23).[40] Fuller's name is thus as inextricably tied to the emergence of nanoscale materials science in the 1980s as is the logo of the IBM corporation. Since 1985, an entire family of "Fullerene" isomers, or "buckyballs," has been discovered and synthesized, all of which contain the same twelve pentagonal com-

Figure 21 Geodesic dome under construction. Note the pentagon, center-left. Courtesy of The Estate of Buckminster R. Fuller.

ponents as C60 along with a variable number of hexagonal faces, including the relatively gargantuan C540. Just as Gerd Binnig and Heinrich Roher shared the 1986 Nobel Prize in Physics for their 1981 invention of the Scanning Tunneling Microscope, Kroto, Curl, and Smalley shared the 1996 Nobel Prize in Chemistry for their discovery of the Buckminsterfullerene.[41] And just as Donald Eigler and Erhard Schweizer's construction of the atomic IBM logo demonstrated the Scanning Tunneling Microscope's capacity to operate not only as an imaging device but also as a means of fabrication, it was Sumio Iijima's 1991 discovery of "helical microtubules of graphitic carbon,"[42] soon to be known as carbon nanotubes, that relayed the explosion of interest in self-assembling carbon structures into an

Figure 22 Buckminster Fuller pointing to pentagonal centerpiece of dome prototype. Black Mountain College, 1948. Courtesy of The Estate of Buckminster R. Fuller.

entirely new branch of nanoscale materials research and fabrication. The domical ends of carbon nanotubes feature the same pentagonal components that allow the Buckminsterfullerene to form a closed cage (Fig. 24).

"The synthesis of molecular carbon structures in the form of C60 and other fullerenes has stimulated intense interest in the structures accessible to graphitic carbon sheets," begins Iijima's article. He continues:

> Here I report the preparation of a new type of finite carbon structure consisting of needle-like tubes. . . . the formation of these needles, ranging from a few to a few tens of nanometers in diameter, suggests that engineering of carbon structures should be possible on scales considerably greater than those relevant to fullerenes.[43]

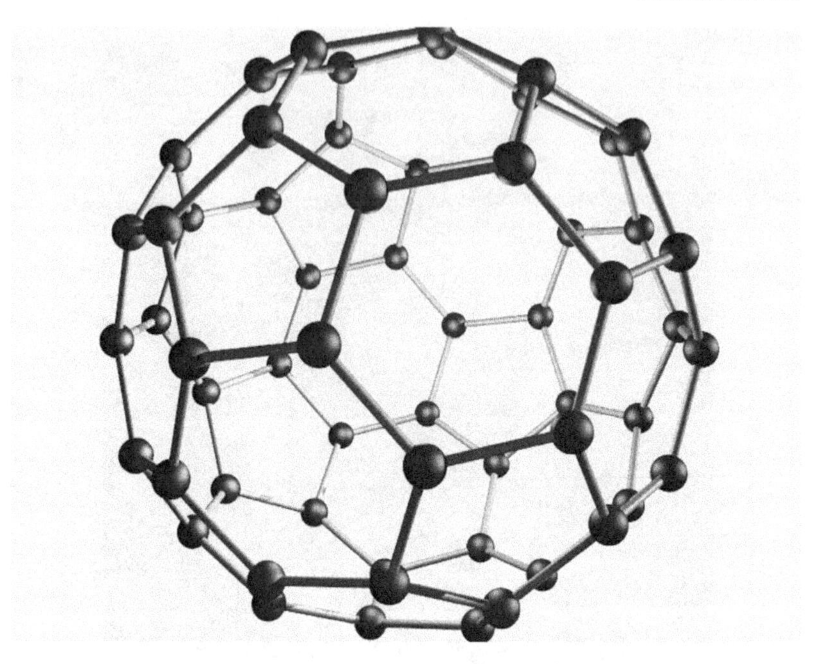

Figure 23 Structure of the Buckminsterfullerene (C60). Courtesy of Creative Commons Share-Alike 3.0.

By July 1992, Pulickel Ajayan and Thomas Ebbesen reported the capacity to mass produce nanotubes with lengths of several micrometers,[44] and nanotubes have since become the poster child of the nanotechnology industry. In 2004 approximately two hundred and fifty tons of single and multiwalled "bucky-tubes"[45] were produced globally.[46]

It is the extraordinary structural efficiency of carbon nanotubes that most obviously validates their attachment to the legacy of Buckminster Fuller's "design science" and his fondness for "ephemeralization": their strength-to-weight ratio is vastly superior to that of steel, and their high flexibility and elasticity make them preferable to traditional carbon fibers, which are stiff and brittle. Whereas most polymer strands can be severed by the cleavage of only a single bond, at least ten to fifty carbon-to-carbon bonds have to be broken in order to rupture a nanotube. In a 2007 review of nanotube research and development published in *Nature*, Pulickel Ajayan and James Tour note that

> unlike traditional micrometer sized reinforcements, such as carbon fibers, nanotubes are truly molecular in size, comparable in lateral dimensions and aspect ratio to poly-

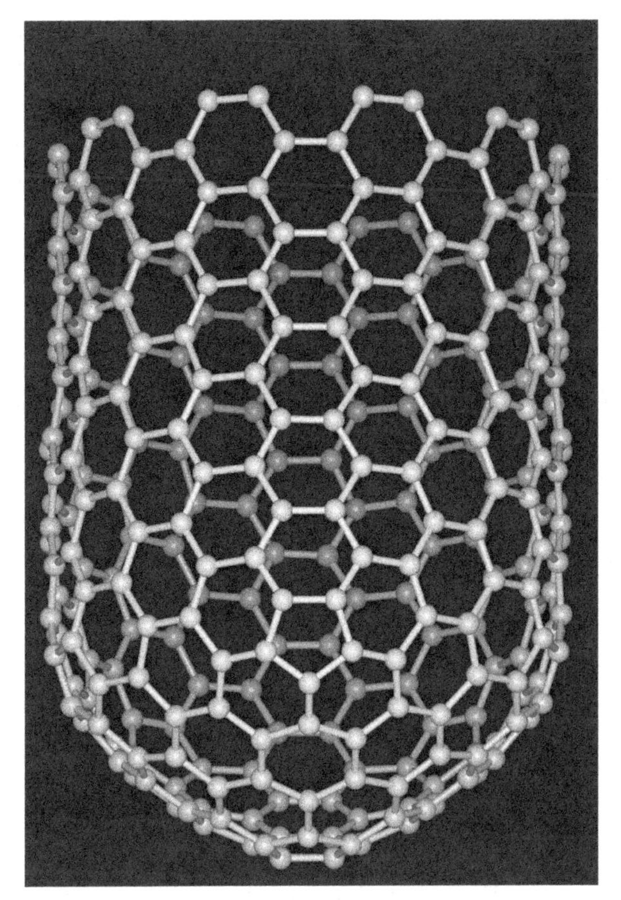

Figure 24 Structure of a capped single-walled carbon nanotube (or "Buckytube"). Note the pentagonal component at lower center, where the structure begins to curve. Courtesy of Creative Commons Share-Alike 3.0.

> mer chains. . . . The skinny carbon-shell structure of nanotubes makes them strong in tension but easily bent, enables them to withstand large strains before failing; and provides a super-low-density framework.[47]

These properties, they argue, may constitute a *limit case* in polymer chemistry: "as the carbon-carbon bond is one of the strongest found in nature, it is unlikely that there will ever be a polymer chain stronger than a nanotube."[48]

It may not be an exaggeration, then, when Harold Kroto claims that the discovery of the discovery of buckyballs and buckytubes "could revolutionize civil and electronic engineering much as steel and aluminum did in earlier times."[49] Others have claimed that fullerenes and nanotubes herald the end of "flat chemistry." Certainly the explosion of chemical research since 1985 involving the use of carbon as a molecular building block bears comparison to the emergence of synthetic chemistry following Kekulé's modeling of the benzene ring in 1865. Boris Yokobson and Luise Couchman have proposed a method for building nanotube tensegrity structures, and Richard Smalley spent the last ten years of his life advocating the global mass production of self-assembling nanowires that might be used to fulfill Fuller's dream of constructing a worldwide energy grid.[50] Smalley is no doubt correct to claim, of the emergence of this new field of chemistry inspired by geodesic design principles, that "Buckminster Fuller would have loved it."[51]

If the construction of the atomic IBM logo exemplifies a particularly iconic mode of nanoscale fabrication (atomic positioning), then the methodology of nanoscale carbon chemistry since the discovery of the Buckminsterfullerene exemplifies a rather different approach to "bottom up" materials engineering: self-assembly. In his contribution to *Buckminster Fuller: Anthology for the New Millennium*, Kroto points out that "our discovery of C60 was most important because it showed that sheet materials on an atomic/molecular scale spontaneously form closed cages."[52] The key phrase here is "spontaneously form." Chemists do not arrange carbon atoms, one by one, into hexagonal and pentagonal networks in order to fabricate fullerenes or nanotubes; they engineer the *conditions* under which particular molecular structures will form themselves. As Geoffrey Ozin and André Arsenault point out, "an atom-by-atom assembly scheme would be hopelessly inefficient: we would likely benefit much more from a chemical self-assembly approach where a strict control of synthetic parameters can build up nanostructures in a highly parallel fashion."[53] This is a *synergetic* approach to materials engineering: one begins with a model of the whole and then induces its component parts to arrange themselves accordingly, without ever addressing them directly. It is this approach to engineering and design that E. J. Applewhite (coauthor of Fuller's *Synergetics*) has in mind when he remarks, in his essay "The Naming of the Buckminsterfullerene," that "there was a greater resonance between C60 and Fuller's writings and design philosophy than the mere congruence of the topology of that molecule and Fuller's geodesic domes." Fuller, he writes, "advanced synergetics as nothing less than a new way of measuring experience and as a new strategy of design science which started with wholes rather than parts."[54]

But one should note how rapidly these two terms—part and whole—tend to reverse themselves within this "strategy of design science." When Fuller wants to explain the principles at work in his domes, he refers us to the "omni-integrity of interaccommodation order" inherent in Nature. His domes are meant to be *part* of that omni-integrity. As Applewhite writes, they are first and foremost *examples* of that integrity: "Fuller did not develop his original great-circle coordinate geometry in order to build domes; he built domes because otherwise people would not understand the geometry."[55] But if the domes are thus *part* of a larger order, when Kroto, Smalley, and Curl attempt to account for the properties of C60 through its geometry the Expo Dome becomes the *whole* to which they refer their polygonal models of carbon bonding. And while the discovery and subsequent mass production of the Buckminsterfullerene exemplifies an approach to fabrication that begins with wholes ("complete" geometries according to which molecules will spontaneously form), fullerenes and nanotubes themselves have subsequently become the parts, or "building blocks," of the more comprehensive project of "self-assembly at all scales," a project of which nanoscale materials science might now be considered a branch.

"It seems that in just a fleeting moment of time," write Ozin and Arsenault in *Nanochemistry*,

> materials self-assembly has changed our entire way of thinking about making new matter. We have seen a shift in materials research from familiar chemistry approaches based on synthesizing molecules, polymers and solids, and engineering physics means of making planar and lateral structures, to a materials self-assembly paradigm, more akin to a lego construction process, where pre-formed and pre-programmed building blocks, with respect to size, shape, and surface functionality, are self-guided or directed to automatically assemble into some sort of device or machine.[56]

Indeed, nanoscale materials science *needs* to be part of such a larger project, since self-assembly promises to provide a "solution to the fabrication of ordered aggregates from components with sizes from nanometers to micrometers."[57] That is, self-assembly is the means of manufacturing whereby the divide between nanoscale entities and "larger, functional ensembles" might be bridged. A major proponent of this perspective, Harvard chemist George Whitesides, points out that "although self-assembly originated in the study of *molecules*, it is a strategy that is, in principle, applicable at all scales."[58] Thus nanoscale fabrication (along with chemistry, robotics and manufacturing, crystallography, microelectronics, work on cellular automata and non-linear thermodynamics, etc.) is taken to be part of a whole called "self-assembly."

Yet self-assembly is itself only one of many approaches to *design*—a catch-all enterprise that, according to Bruce Mau, "is evolving from its position of relative insignificance within business (and the larger envelope of nature), to become the biggest project of all."[59] Mau's formulation begs the question: even bigger than the "larger envelope of nature"? The ambiguity of Mau's statement is emblematic of an ideology in which "design" is both alpha and omega, at once enfolded within and uncannily enfolding the comfortably nested spheres of nature and business, making it all run smoothly. This is the concentric cosmos of "design" (Fig. 25).

The stated aim of Mau's Institute Without Boundaries is to "produce a new breed of designer, one who is, in the words of R. Buckminster Fuller, a 'synthesis of artist, inventor, mechanic, objective economist, and evolutionary strategist.'" For Mau, the role of the nanotechnologist—an instance of this new breed—is to transform materials themselves into "designers." "Materiality," he writes, "has traditionally been something to which design is applied. But new methods in the fields of nanotechnology have rendered material as the object of design development. Instead of designing a thing, we design a designing thing."[60] Insofar as they are both able to do that, "nanotechnology" and "nature" are evidently synonymous. The rhetorical performance of this equivalence is what a student of nanochemistry will encounter in Ozin and Arsenault's introductory textbook, *Nanochemistry*:

> The real truth is that there are in fact nanomachines all around us. They are in the plants we grow, in our pets, and yes, they can be found in almost every cell in our bodies! We are not speaking of a secret government conspiracy to control our minds, but creations of the first and last nanotechnologist, Nature. Over millions of years of evolutionary refinement, Nature has crafted nanomachines capable of much more than our synthetic achievements, even viewed through the lens of our wildest dreams.[61]

This passage is by no means a rhetorical anomaly; it is the standard discourse of what Colin Milburn calls "nanovision."[62] Nanotechnology, to appropriate Helen Vendler's appropriation of Wallace Stevens, is supposedly "part of nature, part of us."[63] If nanoscale materials science entails the project of "designing the molecular world"[64] such that the molecular world designs itself, then Nature must be the first and last nanotechnologist, since "she" is able to design a designing thing. According to this discourse, nanotechnology is natural, since it is already to be found all around and within us, "crafted" over millions of years of evolutionary refinement. Nature and nanotechnology include one another, *by design*.

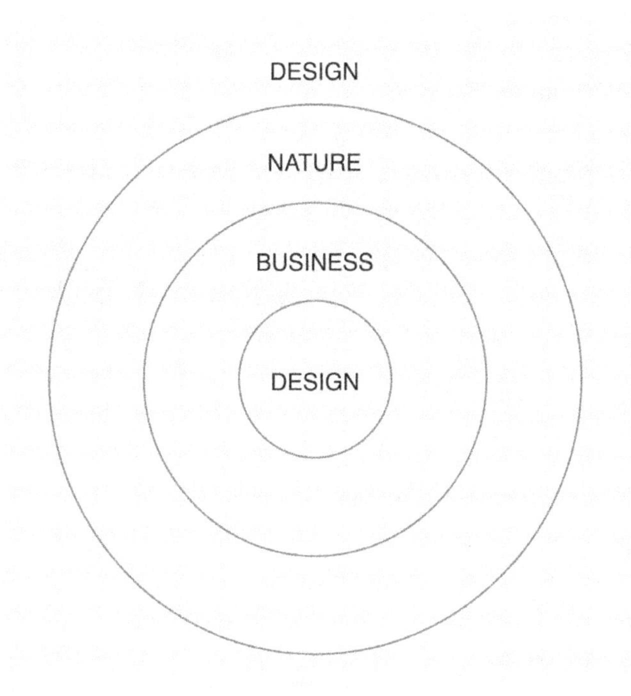

Figure 25 The concentric cosmos of "design."

Design is thus the name of a teleological project, alpha and omega, at once the largest and the smallest "envelope," of all: that within which nature, nanotechnology, and business, via self-assembly, all fall into place. We are told that they do so because, by design, *everything fits*, and every element is where "it wants to be." By design, prophesies the section of Bruce Mau's *Massive Change* devoted to materials science, "we will build intelligence into materials and liberate form from matter."[65] The latter half of this idealist prescription is precisely the ideology passed down to nanoscale self-assembly from Fuller's design science via the "perfect symmetry" of the Buckminsterfullerene.

"BUCKY/FULLERWORD"

In 1985, while Kroto, Smalley, and Curl were modeling the structure of the C60 molecule, Ronald Johnson was planning the third and final book, *The Ramparts*, of his long poem *ARK*. That same year, while at work on an "Ark Essay" expounding his compositional principles, Johnson wrote in his notebook:

Everything means also
something else
(as part of "rhyme"

look up Bucky Fuller's
synergetics, etc.[66]

In his notebook for 1984—the same year in which he declared that his book *RADI OS* was to form a Dymaxion Dome over the whole of *ARK*—Johnson wrote,

Bucky: partition fired partition
fraction ^ wave through fraction
reaction solve reaction.[67]

And on the previous page of the same notebook:

BRAIN

memory abode stem

were epileptics know
as genius because
that storm in the
head realigned all synapse. This *is*
creation

activity

actXity! (a Bucky
Fullerword)

contour.[68]

These "rhymes" and this "Bucky Fullerword" will then constitute the conclusion of "ARK 77, *Arches* XI":

actXity sunder brainstem,
storm in the head
contour everything believable

"fraction wave through fraction,
reaction solve reaction"
inVerse salvation

"ActXity" (evidently: thought-activity under something like the condition of epilepsy, a "storm in the head") functions as or results in a kind of *contour*. It bends that which is believable or aligns it along an *arc*.[69] Through or along such an arc—the bending of an explosive force into a contour—a sundered brainstem, a cognitive crisis, is repaired by the solution of an equation. Fractions are factored, reduced, and canceled out. Every actXion has its equal and opposite reactXion. And this principle of inversion, one term or one stanza answering another, is a salvation, inVerse. As in an ideal tensegrity structure, all conflicting forces find their proper balance in a perfect equilibrium, distributed along an arc.

Johnson acknowledges the inspiration for these compositional principles by dedicating "ARK 53, *Starspire*," to "Bucky Fuller," a personal friend of Johnson's, as he was of John Cage and Ezra Pound, as well as Johnson's friends Hugh Kenner and Guy Davenport. Composed around the time of the notebook entries quoted above, the poem concludes with the following imperatives:

Remove above.
Vault earth
devised,

at once
announce full
Arcady.

***
**
*

Near the center of *ARK*, "over the whole" of which it also presides, the arc of Fuller's geodesic dome functions as a veritable *paradiso terrestre*. What Pound could not accomplish in poetry (according to Johnson, as we will see), Fuller has accomplished architecturally. No longer beholden to the celestial realm toward which the sage aspires on the cover of Johnson's 1969 collection, *Valley of the Many-Colored Grasses*, Fuller has brought the vault of the heavens, the music of the spheres, down to earth in the form of great circles, so that we can "at once / announce full / Arcady" (Fig. 26). Johnson's ode to Fuller suggests that rather than reaching desperately beyond our mundane shell, we can simply "remove above,"[70] since we have at last "devised" a sublunary paradise sufficient unto itself by constructing our own Platonic form: a "domed horizon / measureless / as Zion" (as Johnson writes earlier in the same poem). The stars themselves are gathered into and inscribed in the triangular form at the end of the poem, the form of which also articulates Fuller's geodesic structures. The triangle closing the poem is an inversion, a rhyme of its opening, which is followed by Henry James's famous description of "a splendid intellect":[71]

*
**
***
"The sight
of a great
suspended,
swinging
crystal,
huge, lucid,
lustrous,
a block
of light,
flashing
back every
impression."

Both the domes and Fuller's intellect, Johnson implies, are universe reflectors, "flashing / back every / impression." They are models of the capacity of perfectly ordered architecture, born of a luminous intellect, to inhere within the order of what is so as to return it to itself without distortion.

STRUCTURE RATHER THAN DIATRIBE

It is instructive to bear in mind Johnson's reverence for Fuller's work when assessing the provenance of the long poem that he insisted was "literally an architecture." Here is Johnson's account of his relation to his predecessors, offered in the 1991 note published as a postscript to the Living Batch edition of *ARK*:

> To spend twenty odd years writing a poem, undeterred by risks and shipwrecks of those before, would seem sheer folly. They stand before me, great obstacles. Pound, only a long afternoon in Venice, waving his cane farewell in sparkling background the canal he associated with the writing of *A Lume Spento* . . . W.C.W. maybe a half-dozen visits to Rutherford, when a student at Columbia, rife with sparky theory for American vernacular. . . . More closely, Zukofsky and Olson, braving new schemes for language—The Minimalist and The Maximus—such opposing poles of influence: parities.[72]

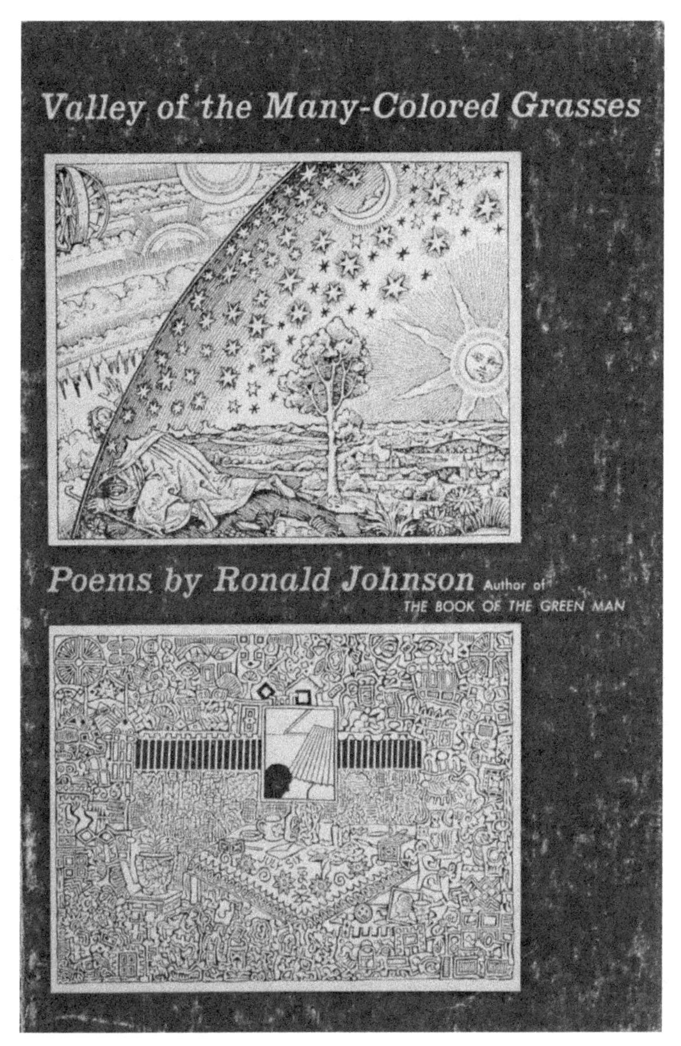

Figure 26 Cover art from Ronald Johnson's *Valley of the Many-Colored Grasses*, 1969.

"I wanted to do a *big* American poem," Johnson acknowledges in his 1995 interview with Peter O'Leary, "always from the beginning it was there."[73] But the works of his forebears in this tradition, with the important exception of Zukofsky's "A," are "shipwrecks": sunken enterprises lacking the solid *Foundations* that Johnson would construct for *ARK* in the first of its three volumes. Conceived of architecturally, it seems the poems of Pound, Williams, and Olson would be condemned buildings: unfinished and structurally unstable "obstacles" whose form and function fall prey to the overreaching ambition of their architects. Johnson "must have begun his work with a worried backward glance at Pound's *Cantos*, Williams's *Paterson*, and Olson's *Maximus*," writes Mark Scroggins:

> These were the looming achievements of the American non-narrative long poem in the twentieth century, and all three of them had petered out into incompletion. Pound sank into depression and silence before he could figure out how to end his *Cantos*; Williams thought he'd finished *Paterson* with its fourth book, only to tack on a fifth seven years later and leave a handful of notes for a sixth at his death; and while Olson specified the one-line poem with which he wanted *The Maximus Poems* to end, the last volume of that work was largely reconstructed by his editors after his death.[74]

The import of both "architecture" and "A," Scroggins argues, is that they offer Johnson a frame within which to *finish* his work. Zukofsky demonstrates that it is possible to complete "a *big* American poem" by operating within a preestablished structure (twenty-four sections) and by recognizing that "the meaningful details that go into the long poem are only as luminous as the overall form in which they are embedded."[75] Johnson's architectural schema for his poem allows him to heed this lesson by making him "able at last to conceive it a structure rather than diatribe, artifact rather than argument."[76] "I had all those big poems in front of me, *A*, and *The Cantos*, and *Maximus*, to look at," Johnson tells Peter O'Leary. "And I said, can I do this any differently? Robert Duncan pointed out that I was the only one to make it an architecture. And when I made it an architecture, then I was settled. Architecture has a plan and an execution and a dedication ceremony."[77] It is his architectural plan that provides Johnson with the structural arc of *ARK*, allowing him to blueprint a complete poem before putting it together.

Johnson routinely cites two "folk" architects as his primary inspirations for the plan of his poem and the method of its composition: Simon Rodia and Ferdinand Cheval. "On his postman's rounds," Johnson writes, "Cheval claimed he kicked a stone one day, then suddenly conceived the idea of building a palace 'like a dream.' . . . Later Simon Rodia's Watts Towers, raising a new realm of mosaic from a Los Angeles slum, gave me a new armature of possibilities."[78] For Johnson, these possibilities present an alternative to the overburdening of the American long poem by "history":

> Olson said that an epic is a poem with history.[79] Zukofsky put a lot of contemporary history and Marxist politics into his poem. William Carlos Williams had a topography,

> a history of all the people around him. . . . But I thought that *ARK* should be like the Watts Towers, like the Ideal Palace of the Facteur Cheval. I wanted it to be *without* history, such that it was constructed of things in my time. It's just filled with snippets: things from books, things on television. When there was a good nature program on, sometimes I got a Rampart or two.[80]

Rodia and Cheval, Johnson writes elsewhere,

> worked from an armature to an outer encrustation of curious rocks, broken colorful tiles or bits of glass, fold and silver foil—anything they could get their hands on from detritus of a world which had not eyes to see. Just so, *ARK* composed itself from the everyday fragments of phrase, words plucked out of context, trouvailles to be worked and knitted and sawn or welded in.[81]

Johnson sees *ARK* as an instance of self-assembly, but it is one that operates according to the treatment of "the materials" prescribed by Creeley ("fused, driven thru, beaten, hammered, fired"). The poem, Johnson says, "composed itself." But it did so out of found materials that had to be worked, knitted, sawn, or welded in. Though Johnson's practice of "scrapture," as he calls it in *ARK* 90, is not exactly akin to Fuller's configuration of refined geometries, the conceptual and practical tension at work here is nonetheless similar to that which we encountered in Fuller's thinking: a tension between whole systems and the materials of which they are composed. *ARK* is *synergetic*, insofar as its self-assembly relies upon the operation of a whole system *into* which the "snippets" of which it is composed can fit. But it assembles itself *out of* contingent materials whose obdurate demands require a manufacturing process, the hands-on labor of knitting, sawing, and welding. That is to say: as will almost any instance of engineering, *ARK* has to rely upon both bottom-up and top-down processes of fabrication. For Johnson, the coordinated articulation of these processes was the work of "twenty odd years."

The tension between the influence of Fuller's ideally synergetic *designs* and the stubborn materialism of Rodia and Cheval's *constructions*, cobbled together according to the contingent demands of whatever bits and pieces were at hand, is operative not only in the way Johnson conceives his poem's assembly but also in the structure into which it is assembled. One might conjecture that *ARK* has ninety-nine sections because the tallest spire of Rodia's Watts Towers is ninety-nine feet high. The spire is that high because Rodia *made it* that high—or, in fact, 99.5 feet—scaling his construction with a window-washing belt and augmenting its dimensions with concrete and scrap reinforcement bars, bent by hand, using the leverage of nearby railway tracks. But while the Watts Towers, Johnson tells us, were a "later" influence, the construction of *ARK* from multiples of tripartite components was a principle decided prior to composition:

> O'Leary: What about the fact that it's in ninety-nine parts. That brings up Dante.
>
> Johnson: Not for me.
>
> O'Leary: Where did the ninety-nine come from?
>
> Johnson: Before I started *ARK*, I knew it was going to be in threes.
>
> O'Leary: That would be the magic number.
>
> Johnson: Three would be the number which carries all the way through. I just decided that was the structure of *ARK*, really, threes.[82]

We might note that Johnson's decision to organize *ARK* into ninety-nine sections comprising three books of thirty-three parts recalls Fuller's devotion to the structural efficiency of the triangle and his passionate advocacy of "the numbers nine and three."[83] The point of noting this is not necessarily to assign direct influence but rather to recognize Fuller's and Johnson's common devotion to a priori principles—numerological commitments—determining the articulation of a structure.

By the time Johnson begins to compose *ARK*'s third book, *The Ramparts*, in the mid-1980s, he builds each of the structurally identical "Arches" constituting that final volume out of tercets, distributing six to a page to form eighteen stanzas over three pages, so that the book's thirty-three sections occupy exactly ninety-nine pages. While the content of those pages may have "composed itself" from "everyday fragments of phrase" and "words plucked out of context" (just as Cheval and Rodia made their work out of "anything they could get their hands on"), the form into which Johnson's fragments are fitted relies upon the modular repetition of structurally identical tripartite components. "I wasn't much thinking of Dante," Johnson insists; "I'm a home grown poet."[84] But if one wants to find an *architectural* rather than a literary precedent for the formal organization of his poem—which would only be to contextualize the structure of *ARK* on Johnson's own terms—one need look no further than the modular, triangular subdivisions of Fuller's geodesic domes, the pure principles of which were laid out in the copy of *Synergetics* to which Johnson was directly attending while planning *The Ramparts*.[85] Architectural tercets. Made in the U.S.A.

VENTS

Synergetics may be a strategy of design science that begins with wholes rather than parts, but Fuller still defines architecture as "the making of macrostructures out of microstructures."[86] It might thus be productive to consider architecture, according to Fuller's definition, as that which mediates between the "opposing poles of influence" to which Johnson assigns Zukofsky and Olson, "The Minimalist and The Maximus." Rodia and Le Facteur de Cheval matter to Johnson because their work illustrates the "edict for either scientist or poet" that he prescribes in his essay "Hurrah for Euphony": "render radiant sense from smallest thing."[87] But this is a lesson that he also learned from another outsider architect.

Conjure lesson
from the
ground up,

he writes in the section of *ARK* dedicated to Fuller,

mortal coil
lock horn
galactic swarm.

It is at the *intersection* of the minimal and the maximal, through the locked horns of the "mortal coil" and the "galactic swarm," that radiant sense is rendered. A "domed horizon" is the sort of structure capable of mediating between these polarities; such a structure is conjured from the ground up but apparently possesses the capacity to render the infinite finite by enveloping it:

Any universe
at all
island enclose
("ARK 53")

Recall Fuller's claim that "tensegrity structures open up completely clear-spanned dome structures of any size." In Johnson's tribute, Fuller's macrostructures are able to enclose the whole, even though they are composed of microstructural parts.

Most readers of Johnson track the procedural elements of his work to the influence of Zukofsky. "It's fairly easy to point out obvious debts Johnson's poetry owes to Zukofsky," notes Mark Scroggins. As he argues, elements of Johnson's work like his excavation of *RADI OS* from *Paradise Lost*, the reduction of the

Psalms into twelve pages of *ARK* in "Beams 21, 22, 23," and the later sections of *ARK* derived from single sources like Thoreau's *Journals* and Protestant hymns ("Arches VIII and Arches XII–XIV") "are all anticipated in a section of 'A'-14 (composed 1964), where Zukofsky boils some nine books of Milton's epic down into six pages of spare verse."[88] In his introduction to Johnson's *Valley of the Many-Colored Grasses*, however, Guy Davenport posits a less obvious precedent to Johnson's treatment of the "snippets" from which *ARK* will be constructed. Citing Wittgenstein's quip that he never wrote a poem because he could never think of a poem to write—and hypothesizing that neither Heraclitus nor Leonardo ever wrote one for the same reason—Davenport wryly notes:

> the miracle by which we have two poems from R. Buckminster Fuller, the Pythagoras *de nos jours*, rises from the transparent fact that Mr. Fuller's Dymaxion prose has twice defied mortal comprehension and has twice been found to be readable when distributed on the page as poetry is, phrase by phrase, or "ventilated," the texts in each instance being word for word the same.[89]

The "ventilated" works to which Davenport refers are Fuller's *No More Secondhand God* (1963) and *Untitled Epic Poem on the History of Industrialization* (1962), the latter of which was published by Johnson's longtime partner, Jonathan Williams.[90] Davenport links Fuller's procedure of "ventilation" to Johnson's compositional method in his poem "When Men Will Lie Down as Gracefully & as Ripe," wherein Johnson "performs for a passage of Emerson's what Buckminster Fuller did for his own prose. He spaces it out and makes a poem of it, though this bald-faced act scarcely answers to the received notion of how a poem is written."[91]

Part IV of "When Men Lie Down," included in *Valley of the Many-Colored Grasses*, opens by quoting a prose passage from Emerson's chapter on Goethe in *Representative Men*. Below this quoted passage, Johnson simply redistributes its language as follows:

NATURE WILL BE

reported.
            All things
are engaged in writing their history.
                                        The air is
full of sounds;
                the sky, of tokens;
the ground is all memoranda & signatures
                                              & every object

covered over with hints
which speak
to the intelligent.
NATURE CONSPIRES.

Whatever
can be thought can be spoken, & still rises for utterance,
though to rude
& stammering
organs.
If they cannot compass it
it works & waits, until
at last it moulds them

to its perfect will,

& is
articulated.[92]

I will return to the content of this poem, crucial to the conceptual underpinnings of *ARK*, later on. For now, we can note that if the procedural technique (altering that content's spatial extension) is derived from Zukofsky's recompositions of source materials, or inspired by Fuller's "ventilated" renderings of his own prose, then the extension itself—the *form* of the poem, the distribution of its "vents"—registers the pervasive influence of open field poetics on Johnson's early work.

Here, then, is an instance in which we can find the distance between Zukofsky and Olson, Minimalist and Maximus, spanned by Fuller, seeker of "min-max limits of simplicity and symmetry."[93] *RADI OS* itself involves only a slight modification of the procedure followed in this earlier poem: the strategic deletion rather than redistribution of Milton's lines in order to achieve a ventilated version of *Paradise Lost* that hybridizes open field and procedural poetics. Given Johnson's characterization of *RADI OS* as a Dymaxion Dome and his later suggestion that it be renamed "Dome Excised Paradise Lost,"[94] it would be hard to avoid following Davenport's lead by looking to Fuller's "poetry," every bit as much as that of Zukofsky, Tom Phillips, or William Blake, as an inspiration for the impetus behind Johnson's ventilated poem. It would also seem that, at least for Johnson, there is a homology between the formal distribution of verbivocovisual forces in open field poetry—its material *lightness* upon the page, the rhythmic extensions that its expansive spacing offers to oral performance—and the dispersed distribution of compressive and tensional forces at work in a tensegrity structure.

CENTRIPETAL/CENTRIFUGAL

Indeed, the most recognizable formal feature of *ARK*—its lineation along a centered column running down the middle of the page—seems itself to have evolved from such an interplay of compressive and tensional forces. In *ARK*, Johnson's lines arrive at a balanced equilibrium emerging from the distributive principles operative in the prosody of his ventilated Emerson. The obvious formal precedent to Johnson's centered lines would be the "spinal" column structuring Michael McClure's work. But it is crucial to attend to the genesis of *ARK*'s lineation in Johnson's earlier poetry. Johnson begins his first long poem, *The Book of the Green Man* (1967), with a clear pledge of allegiance to Olson's breath poetics, understood here in precisely the organicist terms that, in the previous chapter, I have argued are untenable:

> The length of
> breath,
> a sequential foliage
>
> firmly planted in
> our veins,
> we stand in our rayed form[95]

For Johnson, breath forms a "sequential foliage" wherein "we stand in our rayed form," photosynthetically converting photons into measured verse. But whereas Olson understands the breath line as an *immanent* measure of a particular object's relation to a larger field of objects, Johnson figures "the length of / breath" not as an immanent calibration of physical relations or movement but as previously "planted in / our veins." There seems to be a phantom sower regulating the origin of the line, the formal index of which is the tethering of Johnson's lines to the left-hand margin of the page throughout *The Book of the Green Man*. In this sense, the volume truly does fit the mold of the traditional English nature poem, eschewing the topological distribution of the line pioneered by Pound and Williams, and with which Johnson had worked in his earlier volume, *A Line of Poetry, A Row of Trees* (1964). In *The Different Musics* (1966–67), however—printed as the closing sequence of *Valley of the Many-Colored Grasses* (1969)—Johnson generates the columnar form that we find in *ARK* through a compression of the tensional forces operative between lines justified along the left *and* the right margins. The title poem of that sequence opens,

> come simultaneously
> across water,
> accumulating fume, spray, the flex of ripple[96]

The lines continue to hold the left-hand side for just over a page, before suddenly reorienting along the opposite margin:

> "An apparent confusion if lived with long enough
> may become orderly". Charles Ives
>
> . . . accumulating,

On the obverse page, this sequential alternation between margins becomes simultaneous, with parallel poems unfolding along the left and right edges:

Sounds come	I put my ear to
to the ear,	the ground, & heard
transformed: pulling themselves up	the blood
	rush to my head
by the boot-straps:	
	thundering where the roots
as if a new critter	drummed
altogether,	& whispered as they grew: new
created	delicacies, new
out of the marvellous canaries	tangle.
of the air.	And the blood whispered *new delicacies, new*
	tangle.
This exquisite & unending cacophony, sweet	
roar upon roar	New extremities, new labyrinth
	& branching, new, inextricable windings.
swelling, out of the silent	A new foliage of sensings:
shell.	sings & sings . . .

Finally, the tensional field of this "new foliage," itself a major leap in the complexity of Johnson's poetics from the "sequential foliage" of *The Book of the Green Man*, is integrated into a central "trunk," around which the lines mimic a circulation of force that eventually, ecstatically, squares itself:

> And night comes opening its arms like smokes to enfold us:
> THE DANCERS!
> Where their feet touch the earth
> an encircling of plume, diaphanous featherings.
> THE DANCERS
> . . . .
> A spreading effulgence!
> A resplendent "hood" of light!
> A choric turbulence, to which the worlds keep time.

Where their feet struck upon the earth
a pungence sprang up: strewing of thyme & of lavender.

And always the full-winged nights advance:
a field of robins erect
their red breasts facing east, into the rising sun.
OUTWARD!

FIRE IN FIRE, A DANCING FLAME, REDOUBLED LIGHT.

* * * * * * *
* * * * * * *
* * * * * * *
* * * * * * *
* * * * * * *
* * * * * * *
* * * * * * *
* * * * * * *
* * * * * * *
* * * * * * *

This block of ink-fire, a field of robins facing the rising sun, is the first piece of "concrete poetry" to be found in Johnson's major volumes, a harbinger of the visual forms with which he will experiment in *Eyes and Objects* and *Songs of the Earth* before taking these even further in the first volume of *ARK*.[97]

In the final poem of *The Different Musics*, "The Unfoldings," Johnson's central column will unfold again into two parallel columns, before the sequence ends with its lines spiraling around an axis at the middle of the page, this time justified on either side of that axis rather than centered upon it:

The eyes At The End
Of the World—
watching, dilating, inward & outward,

two great stemmed eyes.

THE NEXT SOUND YOU HEAR
WILL BE THAT OF TWO GALAXIES,
EACH
THE SIZE OF OUR OWN
MILKY WAY,
COLLIDING

IN SPACE, 500,000 LIGHT YEARS

AWAY . . .

Centripetal,

Centrifugal:

fugue, & petal.[98]

The overheard galactic collision anticipated here, Johnson discloses, is one of those "snippets" overheard on the radio during composition.[99]

In *The Different Musics*, the strain of the centripetal and centrifugal forces at work in the universe is registered by the fusion and fission of lines into and out of a central column. But in *ARK*, as in Fuller's domes (or in his Dymaxion House, in which compressive forces are distributed through tensional wires suspending a hexagonal living space from a central "mast"), that antagonism of discrepant forces will finally find a perfect equilibrium. Firmly established at the opening of *ARK*,

Over the rim

body of earth rays exit sun

rest full velocity to eastward pinwheeled in a sparrow's

eye

("BEAM 1")

Johnson's centered lines running down the middle of the page are occasionally displaced in *The Foundations* by visual images or passages of prose. But there is only one such interruption in *The Spires*, and in its final volume, *The Ramparts*, *ARK* locks into a perfectly symmetrical system of tercets, a repeating pattern of threes, sixes, and nines.

MATTER: WRITER

If we thus note an increasing degree of formal homogeneity in *ARK*, we also have to recognize the incredible ingenuity of Johnson's concrete poetry. And if we want to correctly diagnose the idealism operative in Johnson's poetics, we need also to observe the active commitment to linguistic materialism that is also manifest in his poetry. "Thinking about thinking moves atoms," he writes in "BEAM 12." Thought, for Johnson, is atomic positioning, and if thinking amounts to moving atoms then it also amounts to moving letters. "matter: writer," he states matter-of-factly ("BEAMS 21, 22, 23"). Thus the figure of the squared circle is not an imagined nonentity but an immediately actualizable configuration of thought-particles:

c i r c
l e c i
r c l e
("BEAM 5")

ARK is "literally an architecture," Johnson insists, not because it is conceived as such or because its books and its sub-subsections are named after parts of buildings, but because the graphemes of which it is composed are microstructures out of which macrostructures are built:

wind Os wind Os
wind Os wind Os
wind Os wind Os
wind Os wind Os
("BEAM 8")

The onomatopoeia at work here, whereby the sound of "w" and "O" and "s" instantiate that of wind, comes into play because the printing of the "O" in uppercase at once registers the drawn-out hollowness of the long vowel sound and renders it a physical form that we *see through.* The uppercase "O," in a relation of spatial proximity to the word "wind," functions as a lettristic synecdoche for "window," presenting the whole through one of its parts and thereby precipitating out the discrepant semantic, visual, and phonemic elements that are normally internal to that word. "Wind" is itself an audio-semantic stereogram, a heterophonic homonym demanding that we register the verb "to wind" within the noun "wind," and also the activity of winding in the circularity of the letter "O" or in the elliptical curvature of an "s." "One seems to see and hear the letters more fully inside this poem than inside most poems," notes Rachel Blau DuPlessis, and "this is a high tribute."[100]

"Particulars evolve," writes Johnson in "BEAM 29":

no where
now here
no where
now here

If we also *hear* "revolve," that is because the concrete poem printed below the phrase "particulars evolve" makes us *see* how to "here" it, how to make a place for words within words by reading forward or backward across the space between them or by reading within them (particulars evolve: love, ars, art, solve, resolve, salve). This is a "quantum poetics," to deploy Daniel Albright's term,[101] insofar as Johnson's particle physics of the word—his decomposition of verbal materials into their protosemantic components—makes legible the paragrammatic agitation that renders any semantic unit irreducible to a simple location or denotative referent, however unconscious of that agitation we may typically be. "No where" and "now here" are two *states* of the "same" word, active in that word at the same time, at once indistinguishable and utterly irreconcilable.

At IBM Almaden Labs in 1993, Don Eigler, Michael Crommie, and Chris Lutz constructed a "quantum corral" using a Scanning Tunneling Microscope to position forty-eight iron atoms in a circular ring. The ripples on the inside of the atomic ring index the density distribution of standing waves of surface electrons, imaged as concentric quantum states trapped within a confined space (Figs. 27 and 28).[102]

In "BEAM 13," Johnson types,

f lux f lux f lux f
lux f lux f lux f l
ux f lux f lux f lu
x f lux f lux f lux
f lux f lux f lux f

The "corral" in this case is the confinement of a repeating set of graphemes, morphemes, and spaces within a square, these units spontaneously forming into diagonal bands as their inscription across horizontal lines is repeated along a vertical axis. The minute separation of the letter "f" captures the fricative's loading of sonic force between the upper teeth and the lower lip, holding it there for the brief instant before the tip of the tongue arrives to flick "lux" across the propagating sound waves. If the poem is performed aloud, these waves will reach their audience *faster than the speed of light*, since the phoneme ("lux") will be heard *just before* we register its denotative value ("illumination"), nested within the word "flux." Read off the page, the photons refracted from the surface of the ink will reach our retina *just before* the brain converts an electrical impulse from the optic nerve into a flow of phonemes. The poem calibrates evidently simultaneous, but in fact discrete, states of sight, sound, and sense as they unfold in language, measuring their complex of interactions through a structure that makes those invisible, inaudible, insensible interactions accessible to the sensorium. Or as Johnson puts it in "ARK 60":

doors of the letters
ring foundry
of this word turn opened
heights within
reveal world

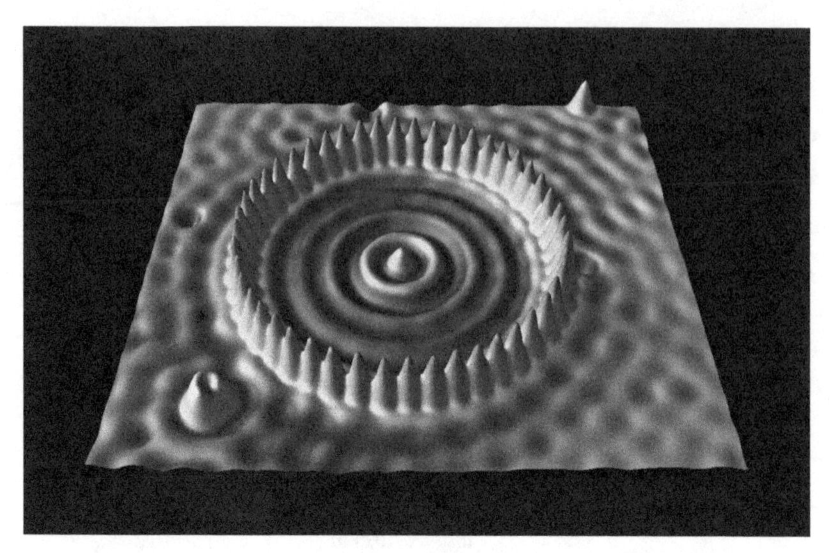

Figure 27 Quantum corral. Atoms positioned and imaged with a Scanning Tunneling Microscope. Image originally created by IBM corporation.

"TO GO INTO THE WORDS AND EXPAND THEM" ("BEAM 28") is the task proper to a practice of poetic fabrication attentive to the physical agency of the letter, and the virtuosity with which *ARK* carries out that task is often stunning.

THE FORESEEN CURVE

ARK not only contains an "ABC Spire" ("ARK 55"), comprised of visual poems for each letter:

n n n n
n ODE n
n ODE n
n n n n
. . .
v a s t
v e i n
v i s e
v o i d
. . .
Z I
O N

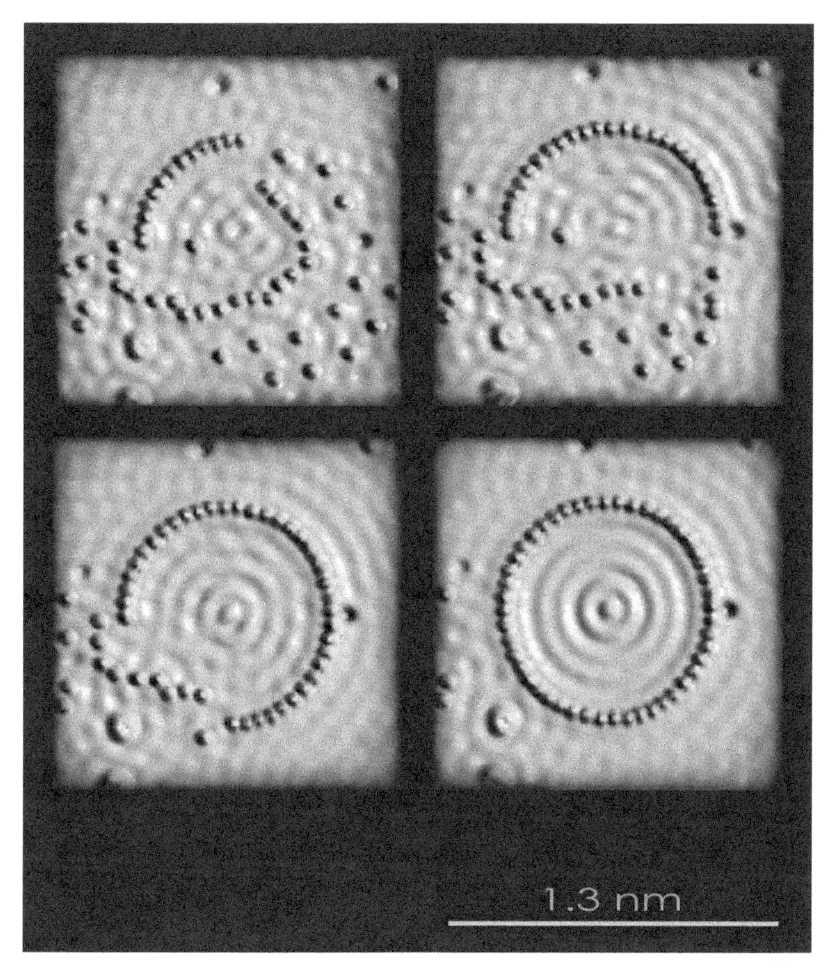

Figure 28 Stages in the construction of a quantum corral. Image originally created by IBM corporation.

It also contains the following index of alphabetic archetypes:

> A is the fulcrum. I, the lever (eye). Out of it ray these three: LFE—single, double, triple vision: L I F E. I's descent from T is the stroke light takes assuming flesh from matter. H weds—is love. When these combine in I they make a windowed quaternity: D closes, J roots, K leafs out. B, P, R, image the female, male, and those reaching between. U contains. C overflows. M, the mountain—V, the valley—W, the wing. O is the Mirror, or

a cosmos made reflective by the hindside of chaos. It is also the egg of S, the instinct's serpent, offering an apple of yinyang everywhere but nowhere to one and all. Z is yellow brick road to question: Quest. Its answer, its obverse, N. (This is the clockwise path.) G, which winds widdershins the sun, is millpool, orworld-pole, Joker. Q is the Unconscious, sperm at the worldegg, positing old meaning to all outward. Y is space. X, time. ("BEAM 28")

Here the graphemes of the Roman alphabet inscribe a network of interrelated signatures, the visual form of each letter evoking and enacting a particular primal scene, form, faculty, process, or principle. Consider the difference between this passage and the concrete poems we have been reading. In a line like "wind Os wind Os," there is no question that we are invited to see the window in "O" via the "o" in "window," which we hear in the line. The letter *functions* as an object, and its function as such is activated by the system of visual, aural, spatial, and semantic relations operative in the line. It is a compellingly ambiguous object, insofar as the line also challenges our immediate association (wind O = "window," therefore look *through* the letter O) through the agency of the verb "to wind," which leads us to the circumference of the circle rather than its center, demanding that we look *at* rather than *through* the letter, that we see ink on paper rather than framed space, and that we register the "sense" of the line as oblique and material rather than immediate and "transparent." But above all, what makes this a concrete poem is that Johnson does *not* write, for example: "O is a window, world-frame, fragile threshold between exterior and interior." In the first case ("wind Os"), the letter functions as an object. In the hypothetical second case ("O is a window"), the object (the letter) operates as a visual symbol. The hard work of *making* that sign material is thus undone by the act of telling us what that material "means."

If Johnson's concrete poetry manifests a commitment to linguistic materialism, his catalog of alphabetic archetypes falls prey to a form of idealism for which, I will argue, we have a precise name, "transcendentalism,"[103] and of which we have already encountered an exemplary formulation:

the ground is all memoranda & signatures
& every object
covered over with hints
which speak
to the intelligent.

If the dictum of objectism is "every sign is an object," the dictum of transcendentalism is "every object is a sign." These are not corresponding principles.

When Dickinson writes, "You there – I – here –" we may notice the manner in which the physicality of the letters is operative in the line: that the grapheme "I" is included in the vertical stroke of "t" and that the detached crossbar of that "t" could function as a dash, "–". But whether or not we think Dickinson was aware of these graphemic effects, what makes the line a powerful instance of constructivist, materialist poetics is that it doesn't inform us (as in Johnson's archetypal alphabet), "I's descent from T is the stroke light takes assuming flesh from matter." The letter "S" is not "the instinct's serpent" but a winding physical form, and when we see it we may or may not utter a sibilant, like a serpent. When Johnson writes:

Z I
O N

We may or may not see the "OZ" (along with "ON" and "IN") in "ZION." If we do, might read the letter "Z" as the winding Yellow Brick Road, *in* Oz, *on* which one walks to the Emerald City. But if we are told "Z is yellow brick road," then the exploration of protosemantic relations is obviated by the predicative attribution of the copula. For a materialist poetics, the value of the letter is precisely that it doesn't "stand for" anything else. For the transcendentalist, on the other hand, every physical form is a "memoranda," a "hint," and a "signature." A signature of what? Of "nature."

"NATURE WILL BE // reported," writes Johnson, transposing Emerson. "NATURE CONSPIRES" to mold "rude organs" to its "perfect will," until, at last, it is "articulated." The difference between objectism and transcendentalism will turn out to be isomorphic with the difference, internal to Black Mountain, between objectism and design science, or objectism and organic form. "It behooves man now not to separate himself too jauntily from any of nature's creatures,"[104] Olson reminds us, because "man is himself an object" as are "those other creations of nature which we may, with no derogation, call objects."[105] But, for Olson, "nature" is not possessed of a perfect will by which it molds objects toward the end of its articulation. Emerson formulates the question of nature in terms evoking the extrapolation from "patterns" of a "form beyond forms" that I have argued is the ur-principle, the constitutive ideology, of both organic form and design science:

> We must trust the perfection of the creation so far as to believe that whatever curiosity the order of things has awakened in our minds, the order of things can satisfy. Every man's condition is a solution in hieroglyphic to those inquiries he would put. He acts

> it as life, before he apprehends it as truth. In like manner, nature is already, in its forms and tendencies, describing its own design. Let us interrogate the great apparition that shines so peacefully around us. Let us inquire, to what end is nature?[106]

The passage by Emerson recomposed by Johnson answers this question: the autotelic end of nature is its own articulation. Every man's condition is a "solution" to the inquiries he would put because nature, through man, answers its own question. When the transcendentalist asks after the end of nature, he asks after the design that it describes. Objectism, on the other hand, begins with the prohibition of this question. "Idealisms of any sort," writes Olson,

> intervene at just the moment they become more than the means they are, are allowed to become ways as end instead of ways *to* end, END, which is never more than this instant, than you on this instant, than you, figuring it out, and acting so. If there is any absolute, it is never more than this one, this instant, in action.[107]

The distance between Olson and Emerson may appear slight, but grasping it depends upon recognizing their radically different conceptions of "nature." For objectism, nature does not have an end subsequent to this instant; that is, an end toward which it "works and waits." This is the point of Olson's rebuke to Robert Duncan in "Against Wisdom as Such." "Any wisdom which gets into any poem," Olson writes, "is solely a quality of the moment of time in which there might happen to be wisdoms."[108] Olson would agree with Emerson that we "act" our condition. But for the former, there is no truth to be apprehended subsequent to that action, only another truth to be enacted. To Olson's "moment of time in which there might happen to be wisdoms," Jed Rasula adds the corollary that "there is no secret archetype or governing *eidos* emanating Idea through the passive plasticity of matter, the mother lode or matrix of manifestation."[109] Objects are creatures of nature; but nature is a detotalized collective of instants and actions, not an "order of things" operating through forms and tendencies to describe a design. The position constitutive of objectism is that nature is not some prior thing that will be "at last" articulated. There are only articulations.

The difference between objectism and transcendentalism is thus the difference and the tension between two different modes of writing we find in *ARK*, the difference between "wind Os" and "O is the Mirror, or a cosmos made reflective by the hindside of chaos." It is the difference between relating to the poem as an object or relating to the poem as the description of a design, the difference between protosemantic instantiation and semantic reference. This distinction is crucial to understanding Johnson's work because it characterizes the tension between the semiotic immanence of his concrete poems (their operation as "signaletic material")[110] and his conflicting commitment to a teleological emanationism that saturates descriptions of natural processes in *ARK*.

ARK is "based on trinities," Johnson asserts, "its cornerstones the eye, the ear, the mind."[111] In *The Foundations*, each of these organs receives, in turn, its adequation to the order of things: "After a long time of light, there began to be eyes, and light began looking with itself" ("BEAM 4"); "Matter delights in music, and became Bach" ("BEAM 7"); "It is as if some eons-old mind (in a time when it could do those things) cast the future on its cold eye, saw Plato's cave, and became our brains" ("BEAM 12"). Behind each of these sentences is the Principle of Sufficient Reason: there are eyes *so that* light may look with itself; Bach was born *because* matter delights in music; human cortical systems are *logical extensions* of eons-old mind. Referring to that "mind," Johnson writes: "where it will look with us . . . is what you and I are doing this instant." According to Johnson's poem there is "no fall of an apple unforeseen" (*ARK* 91). Actions and instants are the correlate of an ancient intelligence with an eye toward the world that we presently embody but which has already awaited us from time immemorial. Self-assembly, in these passages, is not the eruption of radically new events and faculties through unpredictable, contingent, contextual processes; it is the telos of nature's auto-articulation describing its own design through forms and tendencies adequate to an order of things that always already exists. Thus, as we track the apparently aleatory trajectory of an "H" among the first three letters of the alphabet before it locks into place at the end of an epochal proper name—

a H b c
a b H c
H a b c
b a c H
("ARK 46")

—are we to understand that "particulars evolve" by design?

The sourcebook for considering connections between twentieth-century poetics, design science, and transcendentalism is Hugh Kenner's *The Pound Era*, a key chapter of which, "Knot and Vortex," is dedicated to tracing heuristic homologies between Fuller's architecture and Pound's poetics, as well as the background of both in Emersonian transcendentalism. It was an important sourcebook for Johnson. "I never am without Hugh Kenner's *The Pound Era*," he reports in "Hurrah for Euphony." "With only that as a map," he advises younger poets, "you could find your way."[112]

The Pound Era would clearly have been an important text for Johnson's appraisal of what Fuller's design science could offer his own poetics, and the Fuller presented by Kenner is an apostle of "patterned energies"—as is Kenner's Pound. Like the knots that Fuller would tie in the air for his lecture audiences (without any rope), for Kenner Pound's VORTEX is "a patterned integrity accessible to the mind; topologically stable; subject to variations of intensity; brought into the domain of the senses by a particular interaction of words." "The vortex is not the water," writes Kenner, "but a patterned integrity made visible by water."[113] Just so, "the poem is not its language" but the patterned integrity made legible by its language.[114] Even if its physical instantiation is different, the patterned integrity made visible by Fuller's Expo Dome is thus *the same* patterned integrity as that made visible by the water bubbles that he watched forming in the wake of a navy ship in 1917, when he decided that "nature did not use pi" and therefore set out to find the coordinate system of nature that *could* account for the formation of a sphere.[115] The transcendentalist principle at work here, Kenner argues, is that "Nature's processes, Emerson affirmed repeatedly, constitute a system of analogies for the mind to incorporate: 'Things admit of being used as symbols because Nature is a symbol, in the whole and in every part.'"[116] For Kenner, for Emerson, for Fuller, and for Johnson, Nature is a patterned integrity.

By the time Johnson was at work on *The Foundations* of *ARK* in the mid-1970s, he could refer not only to Kenner's investigations of design science in *The Pound Era* (1971) but also to Kenner's *Bucky: A Guided Tour of Buckminster Fuller* (1973) as well as Kenner's *Geodesic Math and How to Use It* (1976), a technical manual that remains a standard reference work on geodesic design procedures.[117] "Most of us," Kenner writes in *Bucky*, "are still unpersuaded that 'materials' are really patterns, that 'things' are insubstantial, that a paradigm for reality is the knot. A rope's very fiber consists of molecular knots, constellated amid vast spaces."[118] On Fuller's behalf, Kenner advocates "the idea of a strength that's designed in, to be distinguished from strength the carpenter hammers in." "Design," he tells us,

> solves problems elegantly, but force imperfectly. Design is weightless; it costs no nails, it costs no wood; it's pure conformity to principle. We do not make "things" out of "materials"; we arrange preferred patterns (and we may learn which patterns work by studying molecules).[119]

Nature and design are one: that is the lesson of the homology Kenner finds between Fuller's design science and transcendentalist philosophy. We can conceptualize the "weightless principles" of tension and compression, Kenner writes in his heady conclusion to *Bucky*, "because we share in the Cosmic Mind in which alone they are real."[120] Kenner's "Cosmic Mind," cribbed from Emerson for a book on Fuller, is also the "eons-old mind" under the auspices of which the particulars of *ARK* evolve.

ALL THINGS PROPORTIONS

"Johnson does not write except to represent the musics of the spheres," deems Rachel Blau DuPlessis, in a critical assessment of his work.[121] In his first book, *A Line of Poetry, A Row of Trees*, Johnson writes,

> There is an exquisite movement, like it were chaos,
>
> but of sweet proportion
> & order:
> the atoms, cells & parsley-ferns
>
> of the universe.[122]

Over twenty years later, in *ARK*, Johnson reiterates this conclusion: "*Proportions*, all things proportions" ("BEAM 31"). Like Fuller's design science, Johnson's poetry is predicated upon the principle that "ratio is all" ("BEAM 25"), and Johnson's poem dedicated to Fuller suggests the latter's geodesic architecture has finally made the design behind the music of the spheres accessible to human engineering. In "ARK 99," the final section of Johnson's long poem, he declares the "music of the spheres solved." Indeed, it is in Johnson's determination to *finish* his epic, to solve it—and thereby to distinguish it from the unfinished "big American poems" of Pound, Williams, and Olson—that his fidelity to the ideology of design is most evident. At the opening of his poem's closing section, he writes:

> Omphalos triumphant
> "only connect"
> end, point of beginning

ARK, then, is both a complete structure and, like the cosmos Johnson imagines, "an organism spirally closed in on itself" ("BEAM 17"). The teleology of design always finds its end in its beginning.

And yet, if the organic integrity and the architectural plan of *ARK* enable its closure into a well-wrought system of ninety-nine parts comprising thirty-three beams, thirty-three spires, and thirty-three arches, it is, ironically, Johnson's handling of its prospective one-hundredth part—its Dymaxion Dome over the whole—that finally clutters the site of its construction. Having initially planned to complete a version of *RADI OS* that would ventilate all twelve books of *Paradise Lost* (rather than only the first four published in 1977), Johnson eventually scrapped this plan, deciding that his first version had "ended when it needed to end."[123] Thus, rather than covering the whole of *ARK*, *RADI OS* (in its unfortunately retitled incarnation as "Dome Excised Paradise Lost") was to be reprinted after the completion of *ARK* as the first part of a volume called *The Outworks*, which would form "a sort of garden surrounding *ARK*."[124] As of yet, however—twenty-one years after the publication of *ARK* and nineteen years after Johnson's death—*The Outworks* has not appeared in print.[125] Taking into consideration this projected but unrealized addendum, one might say that Johnson's poem comprises neither ninety-nine nor one hundred sections but rather something like 99.5, the height of the tallest among the Watts Towers. But one could be even more precise, calculating that Johnson's four "Dome Excised" books of *Paradise Lost* make up exactly one-third of Milton's epic. If we thus consider the reissued version of *RADI OS* published by Flood Editions in 2001 as what remains of *The Outworks*, and therefore as the *remainder* of *ARK*, we are left with 99.333333333. . . . Once again we have the number three, "all the way through," which Johnson "just decided" to make the modular building block of his poem. So again we are returned, via the ghostly geodesic dome on the periphery of Johnson's completed *ARK*, to Fuller's incomplete effort to raise his first geodesic dome and to his first lecture on design science at Black Mountain College in the summer of 1948, drawing the hypotenuse across the diagonal of a square and extolling the virtues of the numbers nine and three.

NATURE POETRY

What is a "nature poem"? That is perhaps the central question with which the trajectory of Johnson's career confronts us, from his early volumes, *A Line of Poetry, A Row of Trees*, *The Book of the Green Man*, and *Valley of the Many-Colored Grasses*, throughout his twenty years at work on *ARK*, and up to his posthumously published book of well-crafted miniatures, *The Shrubberies*. Though perhaps it is a somewhat counterintuitive answer, I have hoped to make evident the answer suggested by *ARK*: a nature poem is the self-assembly of a structure along a foreseen curve, a telos, the *fitting* of material contingencies within a universal design, conceived as a total form. Such a definition might also serve for Fuller's design science, or for the technoscientific mimicry of those "creations of the first and last nanotechnologist, Nature." "In its complexities of design integrity," declares Fuller, "the Universe is technology." If Fuller and the authors of *Nanochemistry* agree that "the technology evolved by man is thus far amateurish compared to the elegance of non-humanly contrived regeneration,"[126] then the goal of such refined technicians as Buckminster Fuller, the fabricators of Buckminsterfullerenes, or Ronald Johnson is to remedy that situation by *making that which we make* adequate to the "design integrity" of Nature. "Design" is thus what presides over and determines the relation between those venerable categories, *technē*, *poiēsis*, *phūsis*. Design regulates the alignment of *craft* and *making* with *nature* via *auto-production*: the self-assembly of parts, qua parts *of* a whole, *into* a whole. Design is a complete poem.

We can therefore posit the following general definition: a nature poem is one that operates under the condition of design. This definition has nothing to do with whether or not a poem is "about" nature. Regardless of what it is about, a poem is a nature poem insofar as it writes the whole qua order, pattern, completion, design. However inorganic the materials they configure or the particles they attend to, design science, nanoscale carbon chemistry, and *ARK* are all "organic poetry" insofar as they endorse and recapitulate a mode of making based on "an intuition of an order, a form beyond forms, in which forms partake, and of which man's creative works are analogies, resemblances, natural allegories."[127] Levertov's definition of organic poetry precisely formulates the identity of nature poetry and design.[128]

Consider Buckminster Fuller at the stern of a navy vessel in 1917, meditating on the principles of energetic geometry at work in the formation of the bubbles that would serve as the model for his geodesic domes:

> I wondered, "to how many decimal places does nature carry out π before she decides that the computation can't be concluded?" Next I wondered, "to how many arbitrary decimal places does nature carry out the transcendental irrational before she decides to say it's a bad job and call it off?" If nature uses π she has to do what we call fudging of her design which means improvising, compromisingly. I thought sympathetically of nature's having to make all those myriad frustrated decisions each time she made a bubble. I didn't see how she managed to formulate the wake of every ship while managing the rest of the universe if she had to make all those decisions. So I said to myself, "I don't think nature uses π. I think she has some other mathematical way of coordinating her undertakings."[129]

Fuller thus expels π from his mathematics. Design science is born of a veritable *horror infiniti*, and the coordinate system of nature upon which it will eventually be based is "a rational, whole-number, low-integer quantation of all the important geometries of experience."[130] The transcendental irrational is the unassimilable Other of transcendentalist design science—or, simply put, of design.

The ideology of "Nature"—in poetry, materials science, or architecture—is thus concomitant with an image of Form that contains and subordinates any given object to the patterned integrity within which it is supposedly included. What Olson calls objectism, on the other hand, is a poetics implicitly aligned with Alfred North Whitehead's postulate that the community of actual things "is an incompletion in process of production."[131] Or, today, we might align such a poetics with Alain Badiou's demonstration in *Being and Event* that Nature, qua whole, is rigorously unthinkable.[132] Perhaps approaches to the relation between *technē* and *poiēsis*, and thus to fabrication, have always been divided by an implicit allegiance either to design or decompletion, to pattern or to the real contingency of what is and what happens, to Nature (with a capital "N") or to what Olson calls "physicality," to "organic form" or objectism. Johnson's fusion of procedural poetics with a commitment to the organic patterning of Nature demonstrates that the real divide in postwar American poetry is not between proceduralism and organic form. Likewise, Fuller's Pythagorean obsession with whole numbers and his rejection of irrational numbers indicate, once again, that the real distinction between approaches to form does not fall between an approach committed to "number" and another committed to the unquantifiable mystery of bodies and psyches. Rather, different approaches to form entail different conceptions of number, and vice versa. The way the organization of bodies is understood is distinguished by a commitment to either the whole within which parts are included or an acknowledgment that the whole inexists.

Recognizing the distinction between *design* and *objectism* thus begs the question not only of contingency but of irrational numbers: what if the structure of *ARK* did not *quite* add up to either 99 or 99.333333333. . . . ? What if that "remainder" were not *exactly* equal to 4/12 of a ventilated *Paradise Lost*, if it were not quite reducible to 1/3, to a reliably predictable decimal expansion—one three after another, ad infinitum—but rather carried an infinite remainder of its own, devoid of any pattern whatever:

3.1415926535897932384626433832795028841971693993751058209749445923078164
06286208998628034825342117067982148086513282306647093844609550582231725
35940812848111745028410270193852110555964462294895493038196442881097566 5
93344612847564823378678316527120190914564856692346034861045432664821339 3
607260249141273724587006606315588174881520920962829254091715364367892590
36001133053054882046652138414695194151160943305727036575959195309218611 7
3819326117931051185480744623799627495673518857527248912279381830119491298
33673362440656643086021394946395224737190702179860943702770539217176293 1
76752384674818467669405132000568127145263560827785771342757789609173637 1
7...

4

SURRATIONAL SOLIDS, SURREALIST LIQUIDS: CRYSTALLOGRAPHY AND BIOTECHNOLOGY IN MATERIALS SCIENCE AND MATERIALIST POETRY

Science gazes at a crystal that promises to answer all questions but that instead captures science with a demand for even more questions. Like the evil genie feared by Descartes or the free spirit loved by Nietzsche, the crystal takes revenge upon the will to truth. The subject tries to solve the object, but meanwhile the object tries to dissolve the subject, and ultimately the object always triumphs.

—Christian Bök, *'Pataphysics: The Poetics of an Imaginary Science*

For that masculine subject of desire, trouble became a scandal with the sudden intrusion, the unanticipated agency, of a female "object" who inexplicably returns the glance, reverses the gaze, and contests the place and authority of the masculine position.

—Judith Butler, *Gender Trouble*

Among the more tantalizing heresies of evolutionary biology, perhaps in all of science, is the speculative thesis of "crystalline ancestry" proposed by the Scottish chemist A. G. Cairns-Smith. "It is proposed," he writes in the opening sentence of his first paper on the subject in 1966, "that life on Earth evolved through natural selection from inorganic crystals."[1] Defending this thesis in his 1982 book *Genetic Takeover and the Mineral Origins of Life*, Cairns-Smith argues that in order to account for the emergence of living organisms with carbon-based DNA-RNA-protein replication sequences, evolutionary theory requires a prior, simpler form of replication that could have served as the "scaffolding" for the gradual assemblage of complex organic macromolecules. He points out that silicon-based crystals, such as those found in ultrafine particles of clay, are excellent candidates for such a scaffolding function, insofar as their structure is copied over each atomic layer of their assembly.[2]

Because inorganic crystals spontaneously develop flaws in their structure that are then replicated over subsequent layers of growth, Cairns-Smith points out that crystal growth involves a principle of variation and the capacity to pass on heritable traits. He stipulates that the transmission of such heritable traits might be considered a form of "genetic information." The theory of crystalline ancestry is based on the possibility that certain physical properties of these differential structures might then constitute a basis for natural selection. In other words, some forms of crystalline organization might be *more* transferable and more conducive to replication than others. These phenotypic variations among crystal structures would therefore have a selective advantage—in terms of the probability of their replication. For example, crystal structures that are relatively stable, yet also somewhat brittle, would propagate more widely in stream beds, where currents would cause such crystals to fragment and travel downstream, replicating their structure elsewhere. A structure that was *too fragile* would disintegrate entirely, while a structure that was *too stable* would resist fragmentation altogether. But a structure that was *somewhat brittle* yet *relatively stable* would break into fragments that would retain their organization as they traveled downstream, replicating that organization elsewhere.

The crux of Cairns-Smith's argument is that the presence of organic molecules, which would bond to these crystal structures, could influence their patterns of propagation and selection. The presence of organic molecules would induce differential patterns of crystal growth, and these molecules might thereby become integral components of replication sequences. Fragments of organic molecules may have aided the formation of structures advantageous to propagation and selection in such a way that the replication of these molecules *itself* became a selective advantage. From the replication of simple amino acids, more complex organic macromolecules—such as nucleotides and lipids—may thus have co-evolved with crystal structures through such processes of selection. This co-evolving replication of inorganic crystals and organic macromolecules would continue until crystal replication was no longer necessary to organic replication. This is the point of what Cairns-Smith calls "genetic takeover." Having gradually co-evolved their own processes of selection and propagation, organic macromolecules would cast off the ancestral crystalline "scaffolding" that enabled their evolution, and more sophisticated biological mechanisms of DNA replication and protein synthesis would eventually develop. According to

Cairns-Smith, "so much had to evolve before our genetic material could have been made that there must have been other genetic material(s) to have supported that evolution." "The present position of nucleic acids in the metabolic scheme," he argues, "can most easily be interpreted as that of an usurper that had come late onto the scene."[3]

Although Cairns-Smith's work is cited approvingly as a promising and ingenious (if partial) theory by both Richard Dawkins and Daniel Dennett, it nevertheless remains speculative and controversial.[4] But regardless of whether it is eventually refuted or proven correct, and whatever lacunae it contains, the theory of crystalline ancestry is nothing if not appropriate to a contemporary scientific context in which the fields upon which it relies and draws together—evolutionary biology and crystallography—have been undergoing their own coevolution over the past seventy years. As evolutionary theory has merged with molecular biology, and as the latter has pursued the reduction of "life" to physical chemistry, it has increasingly been the case that the efforts of molecular biologists to understand the structure and function of biological macromolecules (like proteins and nucleic acids) have relied upon techniques borrowed from the study of inorganic crystals. The emergence of molecular biology in the 1930s was substantially enabled by X-ray crystallography: a technique for determining the position of atoms within crystal structures by passing X-rays through them and analyzing the diffraction patterns caused by the interaction of radiation with the regularly organized electrons in a crystal's periodic lattices. In order to apply this sort of structural analysis to biological molecules, they first have to undergo crystallization. That is: organic molecules have to be grown into regular crystal arrays so that their particular structural and functional characteristics, as *organic* molecules, can be determined.[5]

So molecular biology's conquest of the physical basis of "life" has demanded the crystallization of the biological and the merger of organic and inorganic chemistry. In order to understand fundamental biological functions, we create and analyze crystal structures. Watson and Crick's determination of the structure of the double helix in 1953 relied upon the crystallographer Rosalind Franklin's diffraction data from her studies of periodically organized DNA fibers.[6] The first proteins were modeled by X-ray crystallography in 1958, and as of April 26, 2016, 118,087 protein structures had been modeled by crystallographers and stored in the online Protein Data Bank.[7] Reflecting upon the modeling of the ribosome in a 2001 editorial titled "Whither Crystallography?" the journal *Nature* noted that "the crystallographer's credo that you cannot understand a protein's function without knowing its structure now pervades all of molecular and cellular biology."[8] From the opposite perspective one might add that since the basic geometrical properties of inorganic crystals were more or less understood by the nineteenth century, it is molecular biology that has provided crystallography with its raison d'être over the course its recent history, effectively expropriating its technical capacities and research programs.

This collusion of molecular biology and crystallography, their chiasmatic coevolution, is of interest not only because of the scientific *knowledge* it produces but also because of the practices of *fabrication* it enables. To say that we cannot understand a protein's function without knowing its structure implies that if we *do* understand its structure then we *can*, on that basis, understand its function. If the discourse of molecular biology thus tends to describe structure as the "key" to function, that might also be taken to imply that we can reproduce a function by replicating a structure. It implies that the operative basis of biological processes ultimately resides in the atomic composition of molecular components, a composition that we can reconstruct and alter, just as we can mimic and refine the tessellation patterns of crystal self-assembly in order to produce solids of superior structural efficiency. Noting that "a grand challenge of condensed matter research is to design and rationally prepare complex solids that have predictable and useful properties,"[9] Harvard chemist Charles Lieber points out that because nanoscale imaging devices such as the Scanning Tunneling Microscope enable *local* atomic resolution of crystal structures (rather than the statistical models obtained via X-ray crystallography), such instruments make it possible "to improve rationally the quality of crystals by [seeing] how local crystallographic order changes with different growing conditions."[10] The epitaxial accretion of crystals in homogenously organized sheets provides a model for the Layer-by-Layer (LbL) self-assembly of designer films and coatings, their properties specified at the scale of their auto-replicating molecular structure.[11] The porous structure of zeolite crystals renders them functional as "molecular sieves" and confinement structures for experiments involving nanoscale particles, while also inspiring the fabrication of "open framework materials" featuring self-interpenetrating molecular networks.[12] Meanwhile, in the expanding field of "bionanotechnology," researchers study the structure and function of biological macromolecules in pursuit of "atom-level engineering and manufacturing using biological precedents for guidance."[13] "Looking to cells," writes David S. Goodsell in his introductory textbook on this biomimetic branch of nanoscale engineering, "we can find atomically precise molecule-sized motors, girders, random-access memory, sensors, and a host of other useful mechanisms, all ready to be harnessed by bionanotechnology."[14] While their counterparts in solid-state materials research analyze the atomic configurations of crystal structures in order to afford them rational improvements, bionanotechnologists gather ever more detailed data (with the help of X-ray crystallography) on the amino acid sequences and folding morphologies of protein molecules in order to "design new proteins with custom conformations and functions."[15]

These efforts at the intersection of biotechnology, solid-state chemistry, and materials science come together in investigations of "biomineralization," the process whereby the crystallization of inorganic solids is catalyzed and sculpted by organisms (as in the case of bones and teeth).[16] The exceptionally durable structure of abalone shell, for example, known as mother-of-pearl, or nacre, is formed when ions gradually crystallize within a template of proteins and lipids secreted by the cells of the abalone. Materials scientists working on biomineralization seek to isolate the particular protein sequences involved in the templating of materials like nacre and then replicate and customize those sequences in order to guide the self-assembly of new and more durable crystalline solids, or "layered nanostructures."[17] Doing so involves a heavy dose of genetic engineering, splicing segments of abalone DNA with *E. coli* so that bacteria farms housed in bioreactors will "manufacture proteins to order."[18] Though the rudiments of their structure are initially derived from a "natural" sequence, these manufactured proteins, notes Janine Benyus in her chapter on materials science in *Biomimicry*, "can be anything the biomimic might imagine—proteins that would nucleate an even harder coating than abalone, or perhaps a thin film of crystals with electrical or optical properties."[19]

What links these various endeavors is their approach to biological and inorganic bodies as sites of, and opportunities for, fabrication. Biological molecules and crystal structures, and the interactions through which they are fused into biominerals, are addressed quite indifferently *as materials*. "Proteins become more than directors or scaffolds," writes Benyus, "they actually *are* the material" that is produced.[20] What matters to engineering projects at the limits of fabrication is not so much the distinction between the living and the non-living, the organic and the inorganic, the vital and the mechanical, but rather working relationships between physical components and the milieu in which they are operative. Operating at scale levels at which there is no essential difference between the components of biological bodies and crystalline solids, nanotechnologists resort to *climatological* descriptions like the following conclusion to Richard Smalley's 1995 speech titled "Nanotechnology and the Next 50 Years":

> We've got to learn how to build machines, materials, and devices with the ultimate finesse that life has always used: atom by atom, on the same nanometer scale as the

> machinery in living cells. But now we've got to learn how to extend this to the dry world. We need to develop nanotechnology both on the wet and dry sides.[21]

Smalley retains an appeal to the "wet" and the "dry" as crude descriptions of the worlds of biotechnology and solid-state chemistry, while heralding the capacity of nanoscale engineering to traverse them and displace their opposition. The wet and the dry are figured as the discrepant but interconnected climates that nanoscale engineering is capable of drawing together, and this is the synthetic climate of technoscience at the limits of fabrication.

Caroline Bergvall's *Goan Atom* (2001) and Christian Bök's *Crystallography* (1994/2003) are the poems of this climate. As does nanoscale materials fabrication, these volumes borrow their formal strategies from molecular biology and crystallography. By juxtaposing them here I mean to index the resonance within contemporary poetry of the disciplinary collusion of crystallography and biotechnology in contemporary technoscience. But I do not mean to equate these very different volumes of poetry. As a work that "predicates itself upon an aesthetic of structural perfection," *Crystallography* is "an act of *lucid writing*" that "misreads the language of poetics through the conceits of geology."[22] As a work that predicates itself upon the principle that "Anybod's body's a dollmine,"[23] *Goan Atom*'s feminist/queer poetics uses the "vulgar potential of dropped consonants and arty franglais"[24] to miswrite the perfectionist ethos of biotechnology. While Bök activates the crystallographic constraints of axial symmetry, epitaxial accretion, and atomic tessellation as the means of an inspired mineralogical mimesis, Bergvall splices surrealist artist Hans Bellmer's articulated dolls with Dolly, the cloned sheep, engineering hybrid tropes with which to take on Big Science, all the while working under the rigorous constraint of giving up restraint entirely:

> NO
> workable pussy
> ever was su
> posed to discharge at will
> all over the factory
> sclamation mark(*GA* 53)

In Richard Smalley's climatological vocabulary, one might say that *Crystallography* belongs to the "dry" world, while *Goan Atom* is decidedly "wet." But I want to argue that what these volumes have in common is their contemporary renovation of *objectism*—a poetics, an approach to fabrication, which displaces the opposition of these supposedly discrepant worlds.

While the evolutionary theory of crystalline ancestry draws together organic and inorganic processes of self-replication, and while genetic engineering operates at the intersection of crystallography and molecular biology, Bök's experimental poetry explores the application of crystal geometry as a model for poetic form. I will argue that Bök's volume exemplifies a mode of poetic writing that is both *subtractive* and *inorganic*. It is *subtractive* insofar as the effort of Bök's writing—through the operation of formal and procedural constraints—is to filter out of his work the expressivity of the organic body and the subjective interiority of poetic "voice." It is *inorganic* insofar as the function of this subtractive operation is not only to filter out the expressive "voice" of the poet but also to displace romantic conceptions of *organic form* associated either with the integral rhythmic coherence of traditional meter or the lyric spontaneity of free verse. Against these models of organic form, Bök sutures poetic form to models of inorganic organization and replication drawn from crystallography.

In *Goan Atom*, Bergvall deploys what I characterize as a *performative* method of objectist poetics. Bergvall's is a poetics of articulation and disarticulation, in which the anagrammatic and combinatory play of her language both mimics and parodies the disintegration of the organism and the reconfiguration of biological components by biotechnology. Her work challenges these operations of contemporary technoscience while also adopting them as a poetics through a practice of compositional *détournement*. Like Olson's, I argue, Bergvall's work is concerned with the relation between "the object" and "objectification." But unlike Olson's work, *Goan Atom* addresses the problem of objectification, and its relation to biotechnology, from a specifically feminist and queer standpoint. We can thus read her work against the masculinism of Olson's writing, while also accounting for the manner in which Bergvall reinvents the poetics of objectism in the twenty-first century.

"THE PROPER TIES OF THE CRYSTAL LINE"

Bök describes his text as a "pataphysical encyclopedia" (*C* 156), referring to the "science of imaginary solutions" inaugurated by Alfred Jarry, relayed by Marcel Duchamp and the French writers of the OuLiPo, and carried forward by Bök's mentors of the Toronto Research Group inaugurated in the 1970s: Steve McCaffery, bpNichol, and Christopher Dewdney. In 1980, in a special issue of the journal *Open Letter* devoted to Pataphysics, Dewdney formulated his own theory of "genetic takeover," which might be understood as a linguistic counterpart to the theory of crystalline ancestry. In his article "Parasite Maintenance,"

Dewdney proposed that "the evolution of language, inextricably bound with the evolution of our consciousness as a species, has diverged from its parallel & dependent status with the human species and has become 'animated.' Much like a model of artificial intelligence, or a robot, it has taken on a life of its own."[25] Here, it is language that is the agent of genetic takeover, co-evolving with Homo sapiens until achieving "evolutionary autonomy"[26] as a "separate intelligence utilizing humans as the neural components in a vast and inconceivable sentience."[27] According to Dewdney, this encompassing sentience far exceeds the limited purchase of human consciousness, which is restricted by a regime of embedded linguistic habits. Dewdney posits, however, that "the specialized use of linguistic inventions by the poet"[28] enables occult access to the autonomous operations of language. "This is privileged information," he writes, and "it places the poet in the same vanguard of research as physics, molecular chemistry, and pure mathematics."[29]

Crystallography might be read as a 'pataphysical fusion of Cairns-Smith's and Dewdney's evolutionary speculations. Bök's poetry operates as if its project were to suture the inhuman autonomy of linguistic intelligence directly to the ancestral, pre-biological evolution of the crystal, thereby *bracketing* the organic enunciation of poetic "voice" through a kind of crystallographic ventriloquy, an asubjective/inorganic articulation of mineralinguistic structure. In Bök's volume, it is the geometric and chemical constraints characterizing the replication of crystal systems that regulate "the specialized use of linguistic inventions" serving as the apparatus—the contextual field and the 'pataphysical model—through which the poet attempts an evasion of those habitual patterns of usage restricting cognitive access to the "vast and inconceivable sentience" of language. If our technoscientific dependence upon crystallography as a means of modeling biological molecules is an eerie portent of the Baudrillardian "revenge of the object,"[30] the uncanny homage of our biotechnical future to the deep past of our crystalline ancestry, then *Crystallography* constitutes a poetic counterpart of that technoscientific chiasmus, writing the object's revenge as its apparent capacity to commandeer linguistic evolution. *Crystallography* pursues an inorganic intertwining of crystalline and linguistic replication for which the "neural components" of the poet are expropriated by a program of self-organization traversing minerology and poetics, a program within which the poet is not so much an authorial subject as an object in the larger field of objects.

Crystallography opens with a "Preliminary Survey" in which the traditionally incantatory invocation of the poetic muse is exchanged for a denotative explication of the linguistic sign (Figs. 29 and 30). The poem on the verso page functions as a definition. Its first sentence, "A crystal is an atomic tessellation," alerts us that this will be a book in which the *particulate* materials of language will exhibit their own self-organizing capacities, structural properties, and chemical dispositions. The second sentence glosses self-assembly as the aleatory emergence of order from the disorder of discrete particles, emphasizing the autonomy of formal organization from intention. Apparently a natural artifact that "rivals / the beauty of machine-tooled objects," and could therefore be mistaken for "the artificial product of a precision / technology," the crystal seems, paradoxically, to be neither naturally nor instrumentally produced but rather *autonomous*. It is neither a part of a larger organic unity nor fashioned by the directed agency of technical artifice; rather, it "assembles itself out of its own constituent / disarray."

On the facing page, the denotative reference of this definition is effectively concretized, as the word "crystals" undergoes a process of anagrammatic crystallization, assembling from an archipelago of letters offering a speculative report on the word's origins. The words "astral salt cast astray" (along with "star," "stac," "last," "ars," and "say") are discernible, but the metalinguistic effect of Bök's concrete poetry is to render the semantic function of these semiological units immanent to their physical organization. The immanence of the "meaning" of these signs to their material disposition on the page renders the *sense* of the letters internal to the imperceptible event of their self-assembly (into the word "crystals") rather than making them dependent upon an external referent. The reader, dutifully making meaning from semantic associations, finds the physical substratum of meaning already making and unmaking itself. A diasporic solution of graphemes precipitates the atomic tessellation of the poem's title while the latter seems to dissolve back into a flow of stray particles. It seems that "meaning" is the epiphenomenon of an organization that is already there, on the page, regardless of whether its being there is observed.

In *Crystallography*, language is continually diffracting into casually recombinant arrays and then tightening into firmer structures (Fig. 31). Here, the rule of crystal self-assembly regulates the order of poetic form: the arrangement of words on the page is periodic, their spatial distribution dictated by interlocking combinatory patterns among letters. Laws of alphabetic bonding dictate that the word "lattice" must traverse the "a" of the word "crystal" and connect with the "e" of its own reiteration while the word "crystal" always intersects with the "a" of "lattice" and connects with the "s" or "c" of another "crystal." Shared letters substitute for shared electrons, occupying the "sweet space / between words / between atoms / in among them / strong bonds."

CRYSTALS

A crystal is an atomic tessellation, a tridimensional
jigsaw puzzle in which every piece is the same shape.

A crystal assembles itself out of its own constituent
disarray: the puzzle puts itself together, each piece
falling as though by chance into its correct location.

A crystal is nothing more
than a breeze blowing sand
into the form of a castle
or a film played backwards
of a window being smashed.

A compound (word) dissolved in a liquid
supercooled under microgravitational
conditions precipitates out of solution
in (alphabetical) order to form crystals
whose structuralistic perfection rivals
the beauty of machine-tooled objects.

An archæologist without any mineralogical
experience
might easily mistake a crystal
for the artificial product of a precision
technology.

A word is a bit of crystal in formation.

12

Figure 29 "Crystals," from Christian Bök, *Crystallography*. Courtesy of Coach House Books.

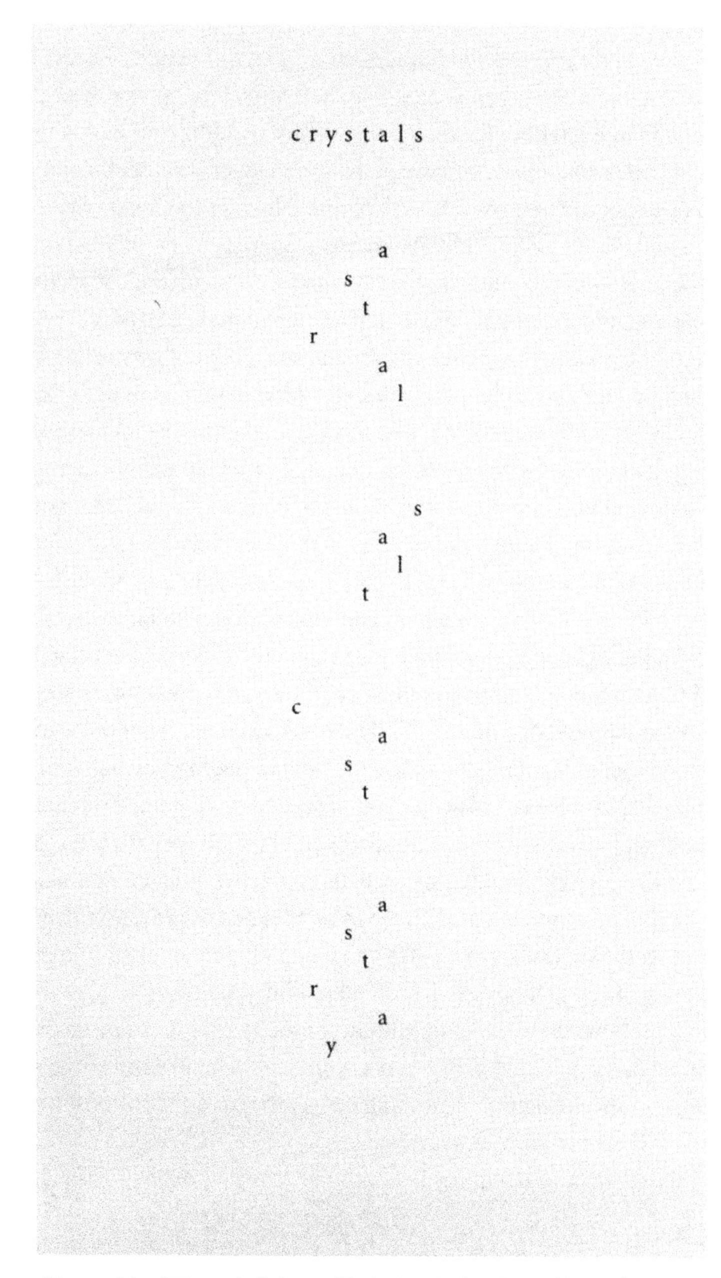

Figure 30 "Crystals," from Christian Bök, *Crystallography*. Courtesy of Coach House Books.

"Crystals partition space with intersecting arrays of parallel / lines," writes Bök, "and these lines, when woven together, form a complex / lattice of letters, used to build trellises for the ivy of thought" (*C* 118). Thus words are grids organized according to the graphemic morphology of their spelling (Fig. 32); snowflakes spell themselves out as they fall (Fig. 33); the letters from which words assemble are revealed by microscopic analysis to be products of crystal replication (Fig. 34); and the "innate crystalline structure" of such letters propagates graphemic fractals (Fig. 35). While the physical analysis of crystalline structure by the crystallographer enables the fabrication of new materials with novel and precisely specified properties, the 'pataphysical analysis of linguistic particles yields novel models of poetic form characterized by immanently constraining rules for the autonomous replication of structural patterns across scale levels. Not content merely to *evoke* a posited reciprocity between word and crystal, Bök sets out to *demonstrate* its 'pataphysical validity. In one instance, he attempts to do so through the algebraic operations of an OuLiPian "mathematical axiology" in which "a letter can become a variable for the value of its position in the alphabet, just as each word can in turn become a relation for the sum of these values."[31] In another instance, the same equivalence is submitted to geometrical proof (Figs. 36 and 37). The word "diamond" refers both to a precious stone consisting of regular octahedrons of crystallized carbon *and* to the geometrical figure that is a plane section of such an octahedral structure. Here this homographic polysemy is effectively concretized: the spatial organization of letters on the page physically constructs the geometrical figure of a diamond, while the proximate evocation of "CRYSTAL: / S(HARD)S" activates a denotative reference to the precious stone.[32] This doubling of the cumulative function of spatially organized graphemes—whose value is at once concrete (◊) and referential ("a precious stone . . .")—itself mirrors the double meaning of the written word ("diamond"), extending and amplifying the ambiguity of the word's meaning into the ambiguity of meaning per se: the uncanny co-constitution of semantic and material qualities of language.

FROM PROJECTIVE TO SUBTRACTIVE

Bök's work contributes not only to the tradition of 'pataphysical research stemming from Jarry but also, more specifically, to a line of geologically oriented poetics traceable from Craig Dworkin's "Treatise on Tectonic Grammar"[33] back through the "rational geomancy" of the Toronto Research Group[34] and Robert Smithson's writing to modernist and midcentury precursors like William

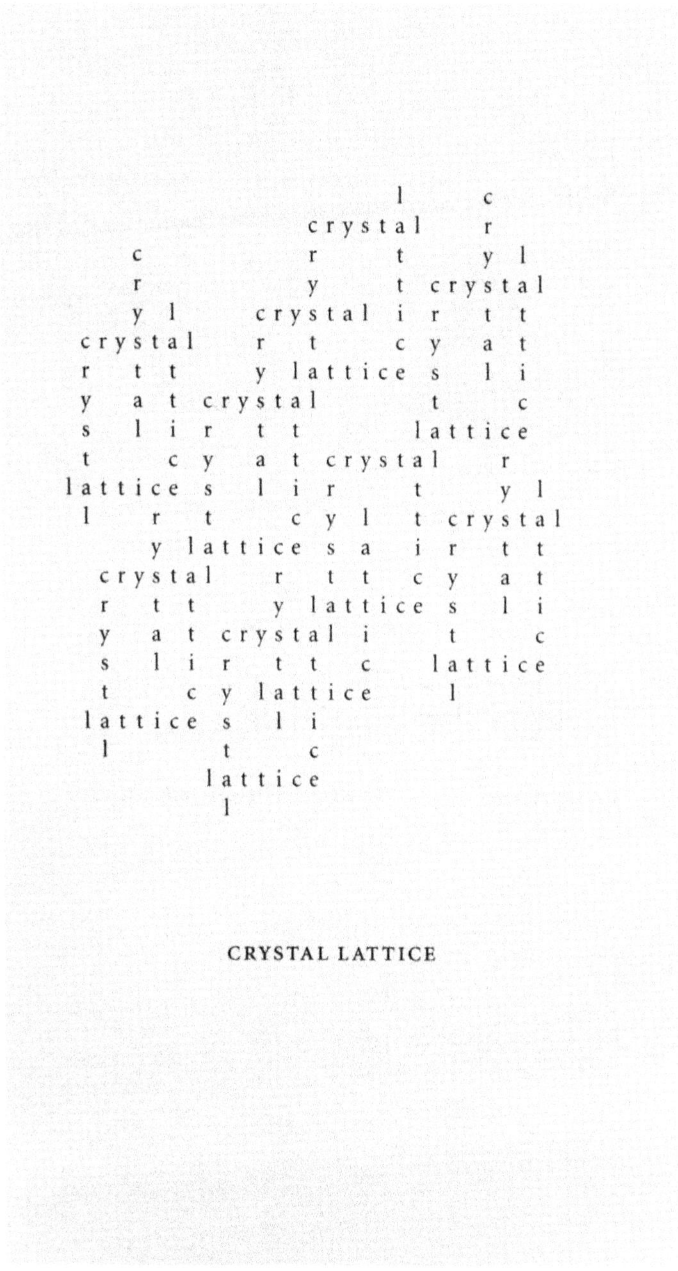

```
                                      l         c
                            c r y s t a l       r
        c                   r         t         y   l
        r                   y         t   c r y s t a l
        y   l         c r y s t a l   i   r     t   t
  c r y s t a l       r     t         c   y     a   t
  r     t   t         y   l a t t i c e   s     l   i
  y   a     t   c r y s t a l             t         c
  s     l   i r       t   t             l a t t i c e
  t         c   y     a   t   c r y s t a l       r
l a t t i c e   s     l   i   r         t         y   l
  l       r     t         c   y   l     t   c r y s t a l
          y   l a t t i c e   s   a     i   r     t   t
    c r y s t a l       r     t   t     c   y     a   t
    r     t   t         y   l a t t i c e   s     l   i
    y     a   t   c r y s t a l   i         t         c
    s     l   i   r     t   t     c       l a t t i c e
    t         c   y   l a t t i c e         l
  l a t t i c e   s     l   i
    l             t         c
                l a t t i c e
                  l
```

CRYSTAL LATTICE

Figure 31 “Crystal Lattice,” from Christian Bök, *Crystallography*. Courtesy of Coach House Books.

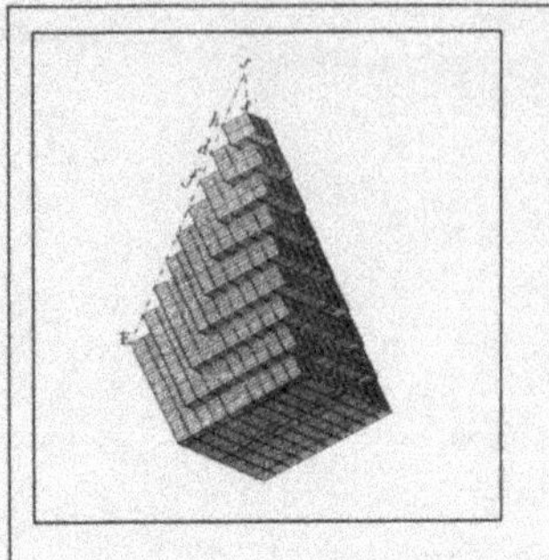

A botanist dropping calcite
shards beholds them break
into regular patterns, every
piece a tiny brick of glass
for building, stack by stack,
apartment blocks of prison
houses – riot cells for souls.

FIGURE 2.0: The polysyndeton of crystal polyhedrons

Lattices form crossword puzzles, diagramless and unsolvable.

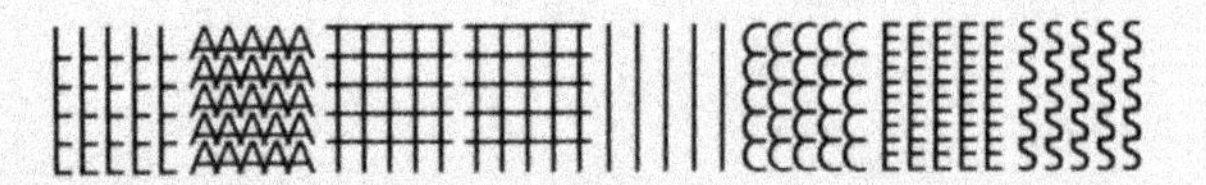

Language acts like a wire grid in a window of reinforced glass.

119

Figure 32 “Lattices,” from Christian Bök, *Crystallography*. Courtesy of Coach House Books.

Figure 33 "Snowflakes," from Christian Bök, *Crystallography*. Courtesy of Coach House Books.

Figure 34 "Photomicrograph of the letter Y," from Christian Bök, *Crystallography*. Courtesy of Coach House Books.

Figure 35 "K-Fractal," from Christian Bök, *Crystallography*. Courtesy of Coach House Books.

12

SIR ISAAC NEWTON
IN THEORY
DEFINED DIAMOND
AS AN 'UNCTUOUS
SUBSTANCE
MUCH COAGULATED'

THE BLOOD
OF ANGELS
CONGEALED
INTO HARD
TEARDROPS
OF PURITY

THE PHYSICIST EVEN HAD
A DOG NAMED DIAMOND

THE GIRL'S BEST FRIEND
WAS ALSO ONCE A MAN'S

LONG DIVISION
AMONG SAWYERS

WORD
= 23 + 15 + 18 + 4
= 60
= 4 + 9 + 1 + 13 + 15 + 14 + 4
= DIAMOND

THE REDUCTION
OF ONE
TO
TWO TO
MAKE MILLIONS

86

Figure 36 "Diamonds," from Christian Bök, *Crystallography*. Courtesy of Coach House Books.

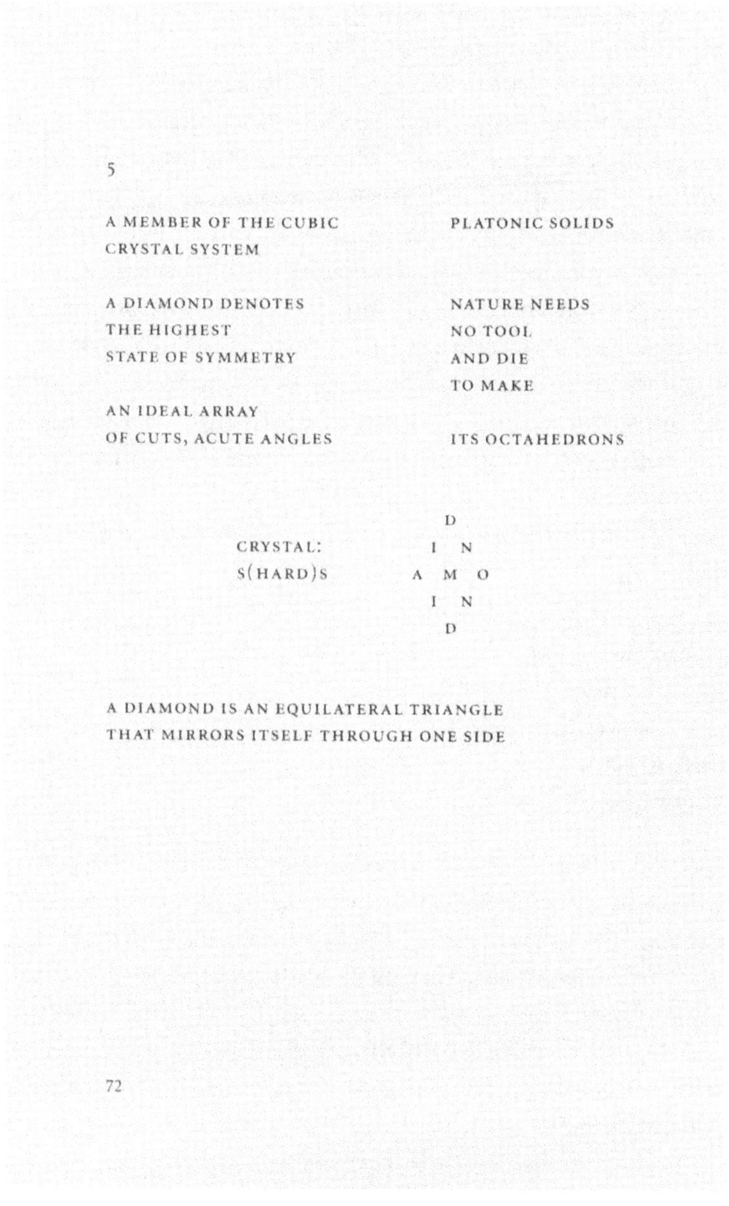

5

A MEMBER OF THE CUBIC
CRYSTAL SYSTEM

A DIAMOND DENOTES
THE HIGHEST
STATE OF SYMMETRY

AN IDEAL ARRAY
OF CUTS, ACUTE ANGLES

PLATONIC SOLIDS

NATURE NEEDS
NO TOOL
AND DIE
TO MAKE

ITS OCTAHEDRONS

CRYSTAL:
S(HARD)S

D
I N
A M O
I N
D

A DIAMOND IS AN EQUILATERAL TRIANGLE
THAT MIRRORS ITSELF THROUGH ONE SIDE

72

Figure 37 "Diamonds," from Christian Bök, *Crystallography*. Courtesy of Coach House Books.

Carlos Williams's *Paterson* and the second volume of Charles Olson's *Maximus Poems*. "Give me geology," Olson says in a series of lectures shortly before his death, "then we don't have to worry about soft human history."[35] In the second volume of the *Maximus Poems* he breaks with the declarative subject position with which the first volume began ("*I Maximus of Gloucester, to You*") in order to excavate the geological undercurrents of that position, "the watered / rock" and the "Carbon Ocean" under the surface of the earth that it indexes.[36] Steve McCaffery's *Carnival* carries on and radicalizes Olson's mobilization of mineral energies rather than humanist values. McCaffery describes the orientation of his major "typestract" poem as "geomorphic," treating linguistics as akin to plate tectonics, activating its "fault lines," cracking open "semanticgeodes," and operating a "realignment of speech, like earth, for purposes of intelligible access to its neglected properties of immanence and non-reference."[37] Christopher Dewdney's contribution to this geolinguistic tradition, his 1973 volume *A Paleozoic Geology of London, Ontario*, deploys the figure of the fossil as a semiological model:

> Devoid of perception the
> blind form of the fossil
> exists post-factum.
> Its movement planetary, tectonic.
> The flesh of these words
> disintegrates.[38]

"THE FOSSIL IS PURE MEMORY," writes Dewdney, as devoid of perception as an engram, a state of semiological immanence requiring no readout. As a radically unintentional inscription or trace of a biological body, the fossil does not index the organism's autonomous agency, its affective expressivity, or its communicative desire. Rather, it indexes the bare fact of its physical existence through the trace of corporeal morphology. McCaffery reads Dewdney's fossil as "ellipse in strata whose extrapolation is the linguistic sign in its state of non-signification."[39] *Crystallography* radicalizes this geological semiology by modeling the autonomy of linguistic production upon the crystal's immanent storage of information through its processes of replication. If a fossil records the residual trace of an absent organism's past existence as information, crystal replication involves the autonomous production of information prior to the existence of any organisms whatever.

If poets like Olson, McCaffery, and Dewdney cite geology as a semiological *figure* of poetic signification, Bök *literalizes* the system of analogy this implies by engaging mineral chemistry as a system of formal constraints regulating the physical construction of the poem. At the outset of their geological turn, the *Maximus Poems* fixate upon carbon as an elemental hinge between the biological and the mineral—between "flower," "protein," and "Diamond (Coal)"[40]—invoking the chemical structure of that element as a common measure. "The carbon of four is the corners," Olson writes.[41] The fact that this line falls at the end of a quatrain, loosely structured around tetrameter rhythm, retroactively constitutes verse form as chemical composition, if only by way of *reference* to chemical properties and the structural parallelism established by such reference. *Crystallography* goes further by suturing language directly to these chemical properties, rendering reference itself structural by deploying the words "for" elemental components in the composition of linguistic molecules (Figs. 38 and 39).

In "Emeralds" the chemical formula for beryl, (Be3Al2(SiO3)6)—the mineral that is called "emerald" when it includes trace amounts of chromium, resulting in a green hue—yields three instances of "BERYLLIUM," two "ALUMINUM," six "SILICON," and eighteen "OXYGEN," all of which interlock on the verso leaf before undergoing proportionally calibrated permutations on the recto. Chromium morphs into "crownland," beryllium into "beaumontage," silicon into "sidereal," oxygen into "opulence," aluminum into "alembic," and so on. The quantity of words in the ode is dictated by the number of elemental atoms in the concrete molecule, while the initial letters of those words are distributed in proportion to those of elemental names. Poems generated from the chemical structure of amethyst, ruby, opal, sapphire, jade, and topaz are interspersed throughout the volume. If, as Bök writes elsewhere, "crystals are acrostics generated by the stochastics of a cage" (*C* 122), and if "a word is a bit of crystal in formation" (*C* 12), then the operative principle of a constraint-based concrete poem like "Emeralds" is that "the properties of the crystalline are proper ties for the crystal line" (*C* 121). In other words, the *formal* properties of crystal geometry furnish Bök's poetry not only with the matter of its *mimesis* (the *imitation* of crystal structure through poetic language) but also with its means of *poiēsis* (the concrete *making* of the poem as material structure) (Fig. 40).

The "inverse law" of Olson's projective verse, of the "artist's act in the larger field of objects," was that "if he stays inside himself, if he is contained within his nature as he is participant in the larger force, he will be able to listen, and his hearing through himself will give him secrets objects share."[42] For Olson, it was thus "the '*body*' of us as object" that enabled participation within the larger field of objects and thereby generated the immanent formation of the poem accord-

EMERALD

```
                        O
                        X
                    O   Y
                  O X Y G E N
                O   Y   E
              O X Y G E N
  C         O   Y   E
  H       O X Y G E N
  R   S     Y   E
  O   I     G   N
  M   L     E       B         S
S I L I C O N     B E R Y L L I U M
  U   C   X   B     R         L
  M   O X Y G E N   Y   S I L I C O N
      N   G   R     L   I     C   X
          E   Y     L   L     O X Y G E N
          N   L   S I L I C O N   G
              L     U   C   X     E
      A L U M I N U M   O X Y G E N
              U         N   G
              M     O       E
                  O X Y G E N
                O   Y
              O X Y G E N
                Y   E
          O X Y G E N
                E
      A L U M I N U M
```

Figure 38 “Emeralds,” from Christian Bök, *Crystallography*. Courtesy of Coach House Books.

crownland beaumontage

sidereal
opulence of sinfulness

opaque, ornate, orphic

silkscreens of silent
orchards

oracular silviculture

alembic of silhouettes

bezels
oblique optics

berylloid observatory

alkali
octane, oxides

ozone overworld of oz

Figure 39 "Emeralds," from Christian Bök, *Crystallography*. Courtesy of Coach House Books.

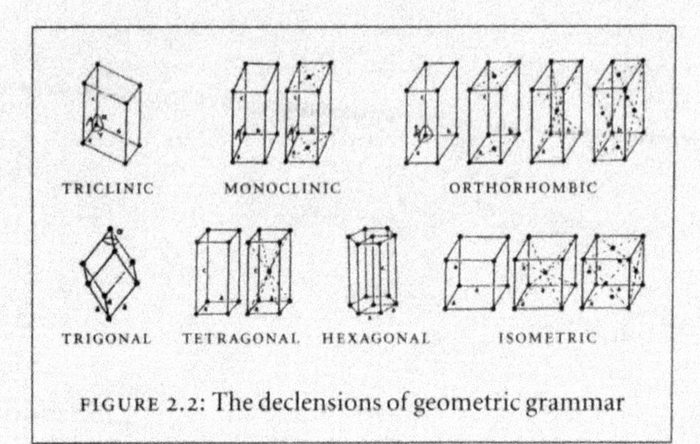

FIGURE 2.2: The declensions of geometric grammar

A T O M C U B E W O R D U N I T Z O N E
T Y P E U P O N O B O E N O D E O K A Y
O P A L B O L D R O S E I D E A N A M E
M E L T E N D S D E E P T E A R E Y E S

Tinkertoys permit children at play to learn through intuition the grammatical fundamentals of a crystalline architecture.

123

Figure 40 "The declensions of geometric grammar," from Christian Bök, *Crystallography*. Courtesy of Coach House Books.

ing to the proprioceptive rhythm of that participation. While Bök's is an objectist poetics insofar as he would seem to agree that composition "is a matter, finally, of OBJECTS,"[43] his objectism differs from Olson's insofar as it is not the body that provides the rhythmic measure of formal coherence in his poetics. The formal law of Bök's constraint-based objectism is not to "stay inside himself," but rather to rely precisely upon those "artificial forms outside himself"[44] that Olson rejects as antithetical to projective verse. Bök's objectism is not *projective* but rather *subtractive*. Olson asks: how can poetic form express the organic functions of the body as a subset of the body's facticity as object? Bök asks: how can we subtract, from the facticity of the body as object, the expressive dimension of its functions as an organism? If the inverse law of projective objectism demands, perhaps counterintuitively, that one inhabit the *specificity* of the body as organism in order to inhabit the specificity of the body as object, then the law of Bök's constraint-based objectism demands that one accede to the *genericity* of the body as object by performing the subtraction of organic expression. The "feeling" of the body—its proprioceptive/affective comportment—is of no moment to the formal machinations of Bök's volume, which demand only that the "neural components" utilized by the "vast and inconceivable sentience" of language become capable of submitting their operations to the "proper ties of the crystal line." Even love, in *Crystallography*, is not the affective experience of an erotically charged organic body but rather "the intricate growth of a crystal / in a dream of time lapse photography" (*C* 102). But if *Crystallography* offers a subtractive rather than a projective objectism, how exactly does this subtractive operation work?

THE AESTHETICS OF THE DEFECT STATE

To grasp the subtractive dimension of *Crystallography*'s inorganic poetics, consider Bök's tribute to the nineteenth-century crystallographer L. A. Necker (Figs. 41 and 42). The Necker Cube is an ambiguous line drawing which, in Bök's characterization, "exists in two dimensions at the same time, while appearing to shift at random back and forth between them without ever seeming to occupy both at once." Here it functions as a figure for the polysemous ambiguity of the homonym described on the facing page. Across the relation between these two pages of the text, the ambiguity of either the cube or the homonym becomes doubled by the ambiguity of the relation between them, as the reader's attention flickers between the contextual domains of geometry and semiotics. If the Necker Cube, as Bök writes, "emulates the form of a photon" in the structural ambiguity that it incubates, the sentence "LIGHT IS NOT HEAVY"—through its double reference to illumination and relative mass—functions as a polysemic linguistic instantiation of the photon's "quantum pun." At the conclusion of the final sentence on the recto leaf, the semiotic stakes of this kind of polysemous play become clear. In this case, it is the syntactical *position* of the word "meaningless" that doubles its meaning; its contextual occupation of a determinate place within the sentence traps the word within the field of its own reference. Because it is "the word at the end of this sentence" the semantic "meaning" of the word recursively cancels itself through its reflexive application: "the word at the end of this sentence is *meaningless*." And because the meaning of the sentence is that the word which concludes it has no meaning, that word becomes meaningless through the self-reference of its meaning. "Delicate words simply dissolve when immersed in their own meaning," Bök writes earlier in *Crystallography* (*C* 50). But in this case, the solubility of semantics precipitates the material obduracy of the word's physicality: because it comes to mean *nothing*, we notice the word's intransigent occupation of physical space, the "meaningless" absorption of light by the black pigment of the ink marking out its constitutive graphemes. The transparency of the word's semantic reference is thus contaminated by the opacity of its material existence. Notably, the *word* does not dissolve, but its *meaning* does, undergoing an auto-subtraction along with the hermeneutic reflex of its reader.

This subtractive register of the poetics of *Crystallography* functions as a trap, albeit one into which only the most careful reader can properly fall. If we notice the manner in which the physical being of the word "meaningless" becomes manifest in the poem I've been reading, what we notice is that our capacity to make meaning, to read signs, has been enlisted by, expropriated into, and then erased by the organization of a semiotic apparatus. The supposedly subjective interpretation of meaning is constrained, captured, and canceled by the operations of a formal system. In the note titled "Lucid Writing" that concludes the volume, Bök explains that

> while the word "crystallography" quite literally means "lucid writing," this book does not concern itself with the transparent transmission of a message (so that, ironically, much of the poetry may seem "opaque"); instead, the book concerns itself with the reflexive operation of its own process (in a manner that might call to mind the surreal poetics of lucid dreaming). (*C* 156)

The relation between crystallography and poetics in *Crystallography* is 'pataphysical rather than analogical insofar as it does not submit crystal geometry and poetic language to comparison by an observer but rather involves these—and the observer—within "the reflexive operation of its own process" in such a way that the observer is at once drawn into and excised from their collusion. *The lucidity of the unliving* is an autonomous reflexivity that, as the theory of crystalline ancestry implies, should be thought as the inorganic medium of replication and reproduction within and through which living organisms come into being. If we think that life is the sort of thing that has "meaning," if we think that life is the medium within which the world becomes significant, the vocation of Bök's subtractive, inorganic poetics is to suggest that, on the contrary, it is the meaningless lucidity of insignificant organization that constitutes the reflexive medium through which the phenomenon called "life" comes to be. "A crystal," Bök writes, "is the flashpoint of a dream intense / enough to purge the eye of its infection, sight" (*C* 37). Far from rewarding a perceptual faith in the phenomenal or hermeneutic capacities of the organic body, the crystal subverts the organism's perceptual faculties, along with its registration and production of significance, while capturing its neurological functions within an autonomous dream of the object's reflexive replication.

The reflexive mechanism by which the volume foregrounds this neurological capture is "enantiomorphosis," the mirroring of crystal forms through an axis of symmetry. Enantiomorphosis is taxonomically applied to language in a "table of crystal systems for classifying letters of the alphabet on the basis of axial symmetry" (Fig. 43). On the previous page, we learn that "a vertical axis makes enantiomers not only of *b* and *d*, but also of *p* and *q*, just as a horizontal axis makes enantiomers not only of *b* and *p*, but also of *d* and *q*." Certain letters are themselves constructed by the reflexive operations of "interpenetrant twinning," such that "*w* takes shape at the moment when *v* twins with its enantiomer through a vertical axis, just as *X* takes shape at the moment when *v* twins with its enantiomer through a horizontal axis." "Such symmetries," writes Bök, "underlie the order of all crystalline forms" (*C* 150). "Even a palindrome," he points out,

> is a kind of enantiomer; for example, the phrase *mirror rim* reveals a sequential symmetry, in which the order of letters in one direction repeats itself when reversed. Each letter is also catoptric in its own structure: the doubled *r*, doubled, the letters *m*, *i*, *o*, each symmetrical through a vertical axis, the gap between the two words a flaw in the gem. (*C* 150)

The position of both the writer and the reader of *Crystallography*—and indeed, the position of the organism in relation to its inorganic poetics—is akin to such a gap: the phenomenal constitution of experience and our hermeneutic con-

L. A. NECKER (1786–1861)

Cubes incubate ambiguity about the rubric of their structure.

Unable to register both meanings of a homonym at the same time, the mind in its indecision must vacillate between two possible, but contrary, interpretations of the same message.

The cube emulates the form of a photon,
a phenomenon existing two ways at once:
both mote and wave – a quantum pun,
in which the universe must take delight.

The homonyms that generate poetic interference split in two when passing through the double-slit experiments of speech.

LIGHT IS NOT HEAVY

English models content to play
with the lead of the cast type in this script.

A man with an appetite for apatite
puts quarts of quartz into barrels of beryls.

126

Figure 41 "L. A. Necker," from Christian Bök, *Crystallography*. Courtesy of Coach House Books.

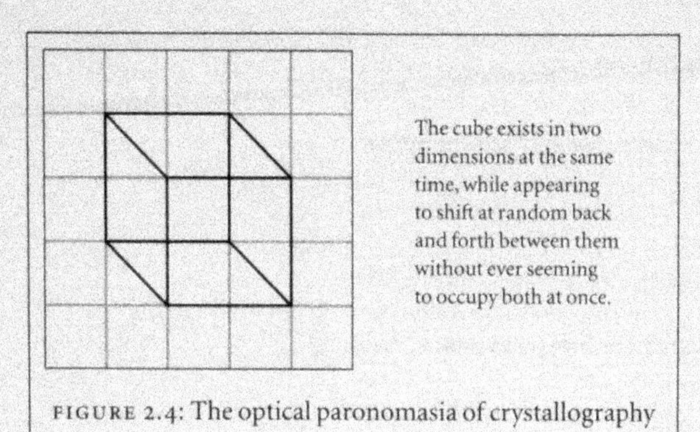

The two words 'two words'
are either 'either' and 'or'
or both 'both' and 'and'
unless one of the words
is two 'words' or one 'two.'

The word at the end of this sentence is meaningless.

127

Figure 42 "L. A. Necker," from Christian Bök, *Crystallography*. Courtesy of Coach House Books.

struction of meaning constitute "a flaw in the gem." On the facing page of the "K-FRACTAL" reproduced above, we encounter the following description of our own position vis-à-vis the enantiomorphic twinning of mineral and linguistic autoreplication:

> When two identical mirrors face each other
> their cycle of self-reflection recedes forever
> into an infinite exchange of self-absorption.
>
> Each mirror
> infects itself
> at every scale
>
> with the virus of its own image.
> Each mirror
> devours itself
> at every point
>
> with the abyss of its own dream.
> When we gaze upon a fractal, we must peer
> at a one-way mirror, unaware of the other
> mirror, standing somewhere far behind us. (*C* 24)

That organism which observes and interprets, whose neural components are seized by language and enthralled by crystal geometry, is an asymmetric flaw in the enantiomorphic twinning of linguistic intelligence and crystal replication, of concept and object, cognition and matter, thinking and being. The subject of science or of literature—observer, writer, reader—looks *at* an image of mineral-linguistic replication (an atomic tessellation, a graphemic fractal), all the while unaware of its position *in* the replicating structure that it looks upon.

This paradoxical position of inclusion/exclusion, wherein the perceiving subject is embodied as a blind spot, is that of the *occlusion*: a foreign substance absorbed into a crystal structure while remaining an exception to its regularity, deviating from and disturbing the propagation of its symmetrical order. The subject who would solve the object is dissolved and absorbed, undergoing a becoming-object as an incongruous element of an encompassing inorganic structure, one in which the flaw or occlusion is itself replicated at each new layer of organization. Rather than lamenting this revenge of the object, Bök welcomes its displacement of organic autonomy and subjective expressivity. He surrenders the organic expressivity of the poetic subject to the inorganic form by which it is occluded. But the fact that such an occlusion *persists* suggests that despite the

CRYSTAL SYSTEM	SYMMETRY AXIS				CRYSTAL STRUCTURE
	DIA	TRI	TET	HEX	
1. TRICLINIC	–	–	–	–	F G J L P Q R
2. MONOCLINIC	1	–	–	–	A M T U V W, B C D E K, N S Z
3. ORTHORHOMBIC	3	–	–	–	H I
4. ISOMETRIC	(6)	4	(3)	–	–
5. TRIGONAL	(3)	1	–	–	Y
6. TETRAGONAL	(4)	–	1	–	O X
7. HEXAGONAL	(6)	–	–	1	–

Unbracketed numerals indicate the mandatory number of axes required for a crystal to occupy a given system. Bracketed numerals indicate the maximum number of optional axes within such a system.

FIGURE 2.8: A table of crystal systems for classifying letters of the alphabet on the basis of axial symmetry

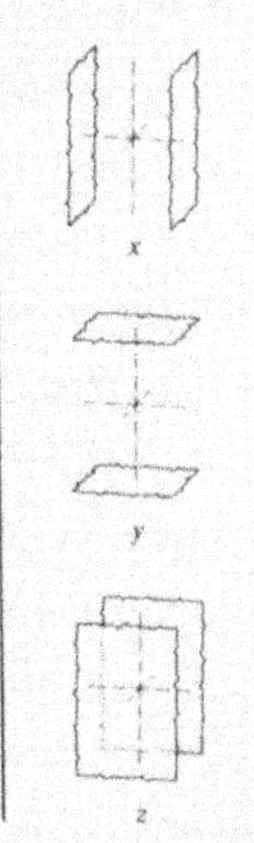

The axis of symmetry describes a mathematical line that passes through the centre of any crystal such that rotation about this line through an arc of 360°/*n* (where *n* ∈ I) causes the crystal to assume a final position congruent with its initial position. For *n* = 1, the crystal can achieve self-congruence by rotating 360° around an *identity* axis; for *n* = 2, the crystal can achieve self-congruence by rotating 180° around a *diad* axis; for *n* = 3, the crystal can achieve self-congruence by rotating 120° around a *triad* axis; for *n* = 4, the crystal can achieve self-congruence by rotating 90° around a *tetrad* axis; and for *n* = 6, the crystal can achieve self-congruence by rotating 60° around a *hexad* axis. No crystal exists with axes of symmetry for *n* = 5 or for *n* > 6.

The number of such axes of symmetry in a given letter determines the crystal system to which the letter belongs: the crystal H, for example, coincides with itself when rotated 180° through any one of three different orientations, and thus the letter belongs to the orthorhombic system, whose members typically have three such diad axes. The alphabet consists predominantly of monoclinic crystals, three types, all having a single diad axis of symmetry: a) three letters symmetrical only through the *x*-axis; b) five letters symmetrical only through the *y*-axis; and c) six letters symmetrical only through the *z*-axis. The triclinic system, the next most common system in the alphabet, contains letters with no axis of symmetry other than their infinite number of identity axes.

The science of crystallography suggests that both the isometric system and the hexagonal system do occur in nature; however, no poet searching throughout the world of language has yet discovered a letter that fits into either system – a mystery that has led some crystallographers to speculate that crystals expressing such a rare degree of symmetry can only exist under the most extreme poetic conditions: perhaps the low temperatures found only in the voids of outer space or the high pressures found only in the cores of neutron stars – conditions difficult for writers to reproduce in the laboratory.

151

Figure 43 "A table of crystal systems for classifying letters of the alphabet on the basis of axial symmetry," from Christian Bök, *Crystallography*. Courtesy of Coach House Books.

"aesthetics of structural perfection" to which *Crystallography* avowedly aspires, and despite the subtractive rigor of its inorganic formalism, there remains a trace of the organic body that intervenes between the autonomy of crystal replication and linguistic intelligence. Even a dream of perfect form must suffer the occlusion of its dreamer (Fig. 44).

If the phenomenon of "life" seems self-evident, while its concept remains unthinkable, perhaps that is because life is an occlusion of the symmetrical relation between thinking and being, that which intervenes between existence and cognition. Writing is the record of that occlusion, by which it is both exposed and effaced. By way of a detour through "life," being writes itself as thought; thought writes itself as being. The medium of this detour, "life" is that occlusion which cannot think the very concept of its existence. Life is the occluded asymmetry that intervenes between thinking and being, and "writing," inscribing this asymmetry, as we read in *Crystallography*, "is the superficial damage endured by one surface when inflicting damage upon the surface of another" (*C* 124).

Bök reconstitutes the *subject* of writing as the *object* of the occlusion: this is the key to understanding the relation of *Crystallography*'s 'pataphysical project to the ideological stakes of contemporary efforts by materials scientists to "improve rationally the structure of crystals" and to model the morphogenesis of synthetic materials on processes of crystal self-assembly. While such nanoscale engineering projects, like Bök's own mineralinguistic engineering, apparently abide by an aesthetics of structural perfection, they too are inevitably confronted with the irreducibility of the flaw, or defect—with an exception they will try to convert into a rule. Consider chemist Geoffrey Ozin's characterization of this fact:

> When introducing solid-state materials chemistry to the student, after having laid the foundation of solid-state synthesis, structure determination by X-ray diffraction, and the electronic band description of solids, I often announce, "*defects, defects, defects, there is no such thing as a perfect crystal, if it did exist it would not be terribly useful and applications would be few and far between.*" Of course, students find this upsetting at first and difficult to comprehend until they learn about the aesthetics of the defect state and that it is the imperfection of solid-state materials rather than perfection that provides them with interesting properties and ultimately their function and utility. "Perfecting imperfection" in solids and knowing which kind of imperfection to perfect to achieve a particular objective is a challenging yet important and universal concept for the student of materials chemistry to grasp.[45]

8

TRANSPARENCY
IS THE GAUGE
OF ALL VALUE
WHEN CUTTING
WHEN WRITING

A DIAMOND IS
AN ABSENCE
OF NOTHING
BUT DARKNESS

OCCLUSIONS
SUPPOSEDLY
DEPRECIATE
PERFECTION

OCCLUSIONS
PERFECTION

78

Figure 44 "Diamonds," from Christian Bök, *Crystallography*. Courtesy of Coach House Books.

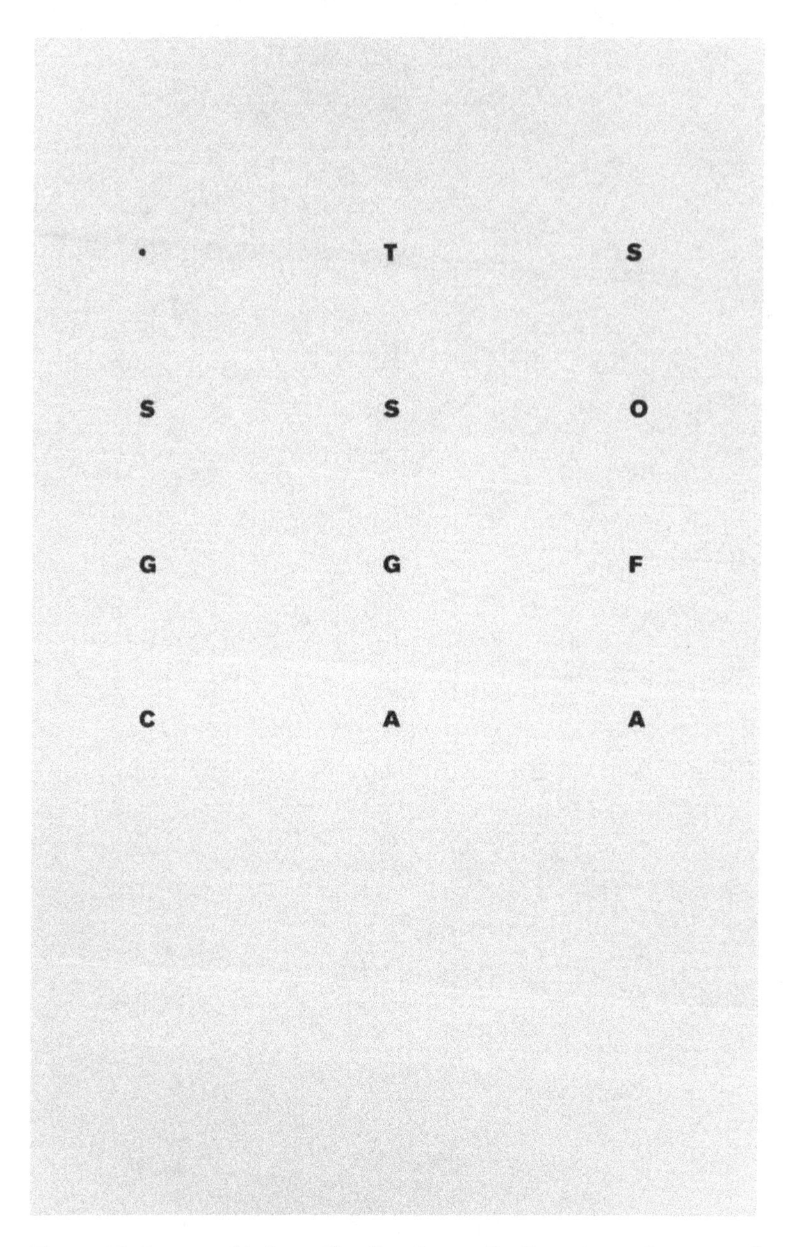

Figure 45 Letter grid, from Caroline Bergvall, *Goan Atom*. Courtesy of Caroline Bergvall.

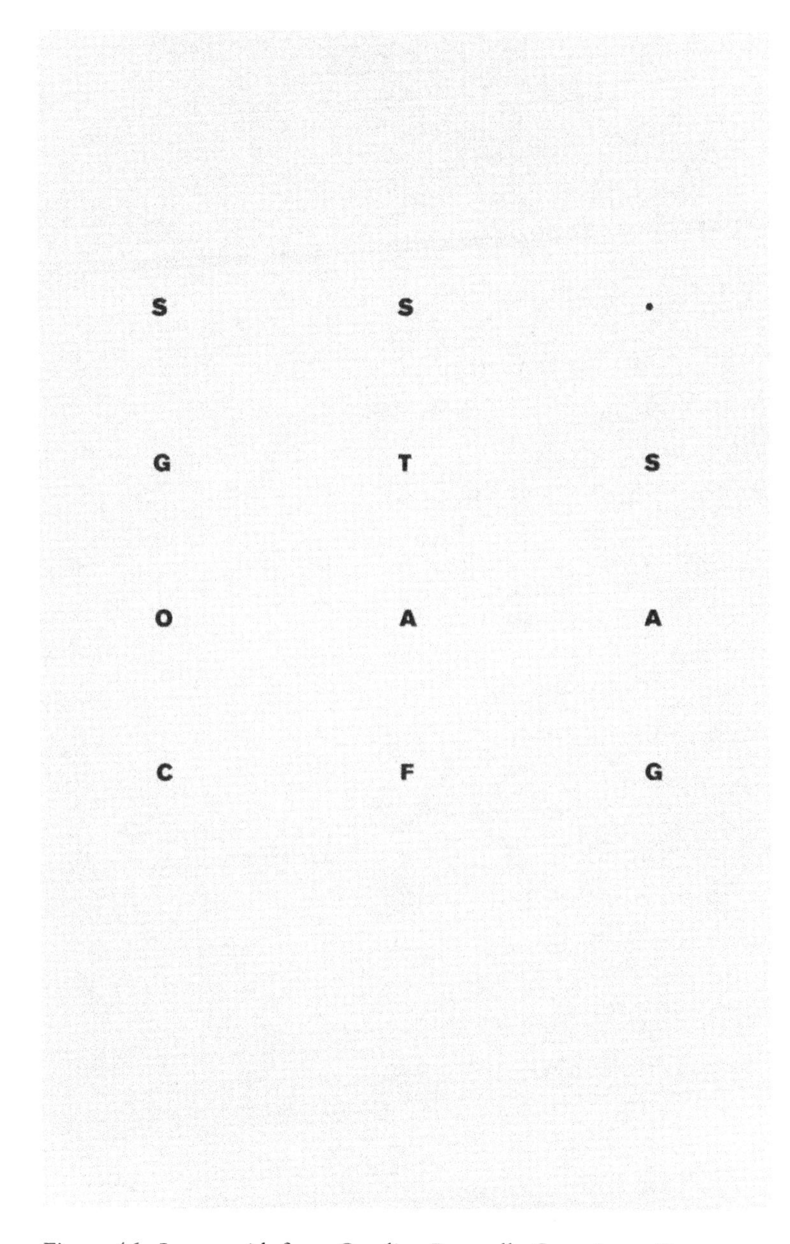

Figure 46 Letter grid, from Caroline Bergvall, *Goan Atom*. Courtesy of Caroline Bergvall.

S

G

O

C

19

Figure 47 “Cogs,” from Caroline Bergvall, *Goan Atom*. Courtesy of Caroline Bergvall.

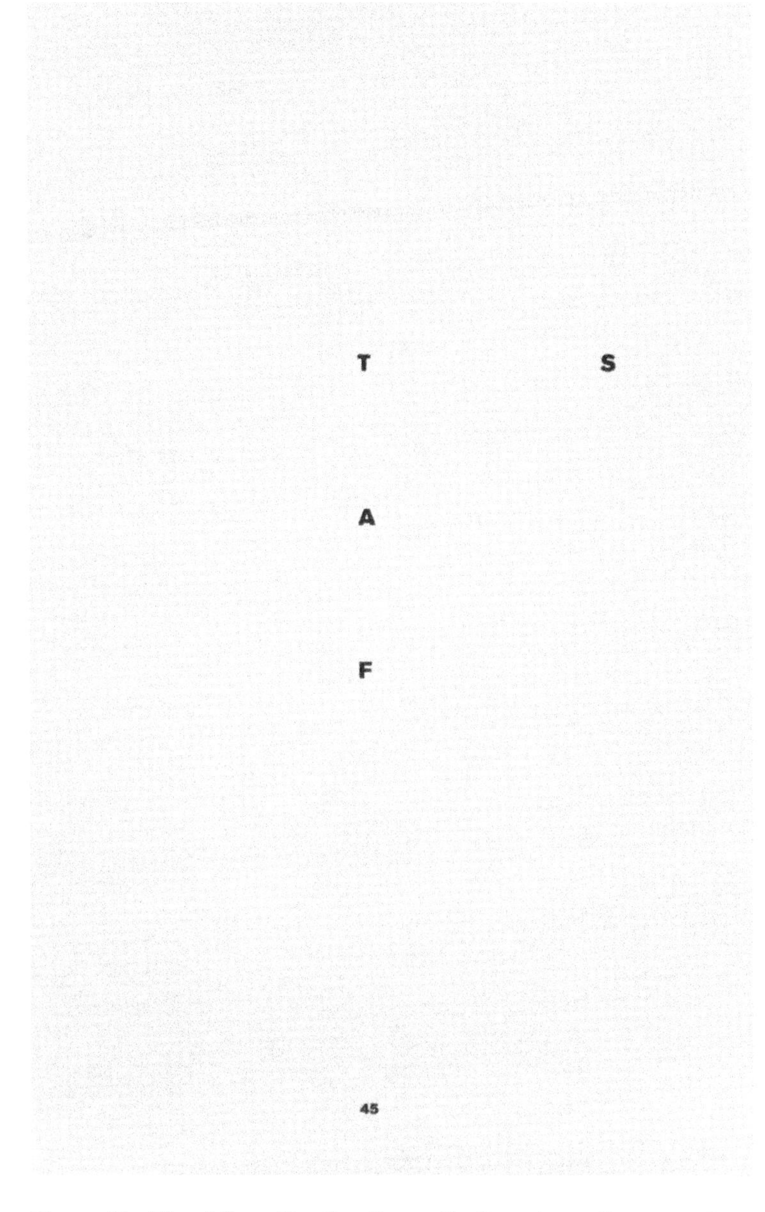

Figure 48 “Fats,” from Caroline Bergvall, *Goan Atom*. Courtesy of Caroline Bergvall.

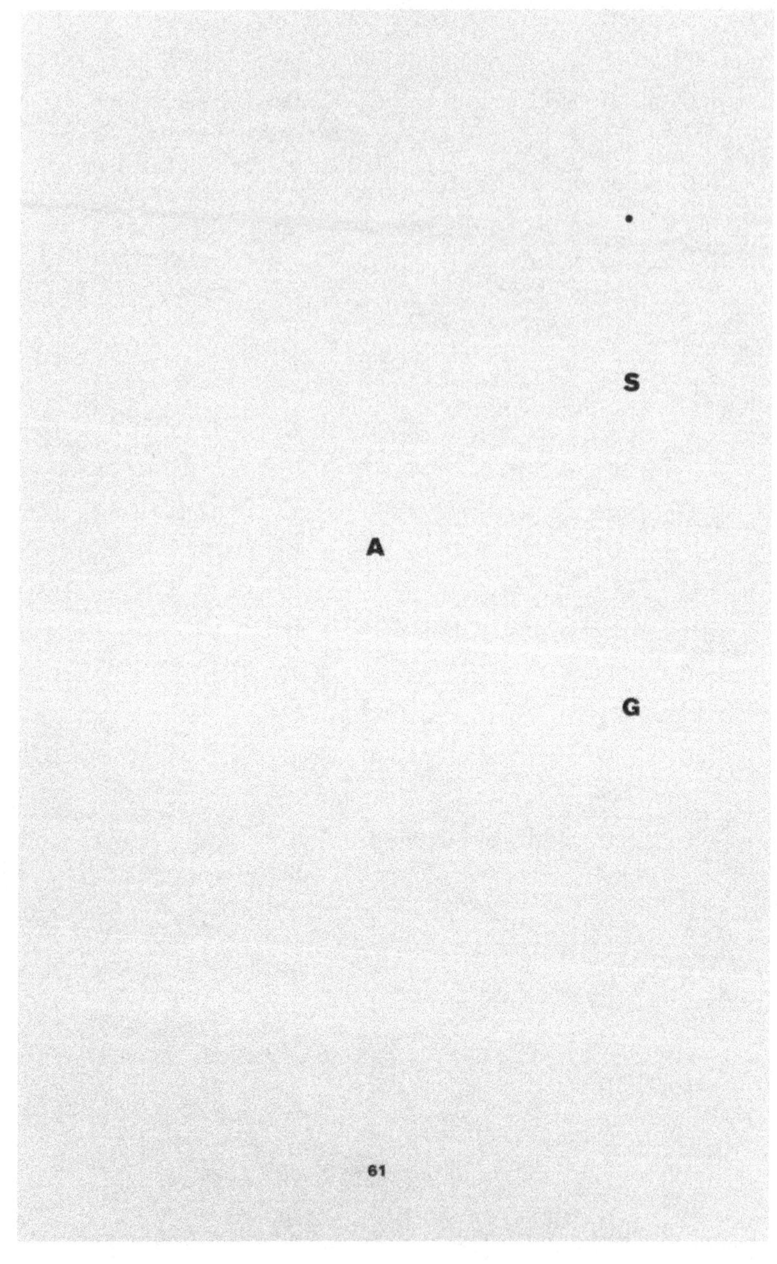

Figure 49 "Gas," from Caroline Bergvall, *Goan Atom.* Courtesy of Caroline Bergvall.

Just as the indeterminacies of quantum phenomena present a constructive opportunity rather than an epistemological aporia for the nanoscale materials engineer, who attempts to productively integrate their effects into multiscalar systems, the imperfection of the crystal is no barrier to the solid-state materials scientist, whose increasingly finite analysis of structural flaws enables a functional deployment of the exception. The "aesthetics of the defect state" is thus the enantiomorph of the "aesthetics of structural perfection," and the advantageous handling of their interpenetrant twinning amounts to "perfecting imperfection." But whereas the subject of science thus practices the rational perfection of the object's imperfection, the surrational premise of *Crystallography* is that the putative subject of writing is in fact the very imperfection of the object that it dreams of perfecting—an occlusion within the interpenetrant twinning of linguistic and crystalline evolution, ideally submitted to, but inevitably deviating from, their formal constraints. "If poetry cannot oppose science by becoming its antonymic extreme," writes Bök, "perhaps poetry can oppose science by becoming its hyperbolic extreme."[46] At the hyperbolic extreme of both technoscience and poetry—that is, at the limits of fabrication—'pataphysics wagers that every perfection of imperfection imperfects itself: that every imaginary solution precipitates an unimagined occlusion.

OBJECTS THAT MATTER

Like *Crystallography*, Bergvall's *Goan Atom* begins with a preliminary survey. On its opening pages, we find a succession of two lettristic grids, printed in a bold sans serif font (Figs. 45 and 46). The abstract order of the grid as a form of organization is mediated and offset by the concrete bulk of the units that constitute it. Like the concrete poetry of Bök's introductory paean to crystalline self-organization, this opening gesture emphasizes that the words we will encounter in the pages thereafter are *composed* of letters and that these letters are themselves physical marks on a page organized and reorganized into arrangements that are not only sequential but also spatial. Like Olson, Bergvall evidently wants us to consider the page as a relational field on which signs are not just written but configured, where their status as differential units depends on their distribution in space. Two pages after these grids we find a list of three numbers, leading us to the title pages of the book's three subsections, COGS, FATS, and GAS (Figs. 47, 48, 49).

Extracted from the second of the two opening grids, the titles are printed according to the position on the page of that grid's discrete units. The suggestion is that the formal structure of the book, its division into titled sections and numbered pages, is inextricable from the spatial arrangement of the material units of which it is composed.

Following the leaf on which we find the page numbers of the book's three sections, this tension of the volume's material components against their formal configuration is reasserted as we encounter an illegible, unintelligible, unpronounceable constellation of inkblots (Fig. 50). If the volume opens by insisting on the fact that graphemes are material units, it also insists that we pay attention to what they are made of. Bergvall's feminist poetics merges with her materialism here, as the menstrual thematics punctuating *Goan Atom* figure these blots as an instance of textual spotting. Offering a sublunary counterpart to the astral diaspora that opens *Crystallography*, the blots index an excess of corporeality—whether scriptural or somatic—over the orthographic or hygienic regimes that would contain or discipline the boundaries of unruly fluids and physicalities. But if we thus begin to understand that Bergvall's volume will deal with the "wet world" of biology and biotechnology, in apparent contradistinction to the solid-state chemistry at stake in *Crystallography*, we should note as well that the appearance of these fluids on the page is also ordered and discrete, involving a topological grammar of its own. These are at once unbounded flows and bounded objects that mark the page in a highly *particulate* configuration, foregrounding the manner in which any liquid is articulated into discrepant components, just as any solid is always involved in a processual dispersion, even if only at a glacial or even cosmological pace.

Two-thirds of the way through *Goan Atom*, we arrive at a passage that decisively formulates the elements of an objectist poetics that I have been drawing out of the opening pages of Bergvall's volume:

> Blt **o** by Bolt
> Every single P
> art is a crown
> to Anatom
> (*GA* 56)

The open inkblot of the boldface letter "**o**" in the first line forms the circumference through which a bolt might be screwed: the sexual mechanics of body parts are so thoroughly implicated in the language of *Goan Atom* that they are discernible at the level of the grapheme. But the larger import of this passage is its formulation of what I want to call Bergvall's *poetics of articulation*. This is a poetics that operates blot by bolt: through minute reconfigurations

Figure 50 Ink blot, from Caroline Bergvall, *Goan Atom*. Courtesy of Caroline Bergvall.

of the particulate units from which physical bodies are assembled, units that are at once solids (bolts) and quanta of flows (blot). Bergvall's is a constructivist poetics in which "art" poses the problem of "part," and vice versa. The co-implication of blots and bolts (stereotypically gendered figures of wet and dry worlds) and their mediation by a detached concrete signifier (the letter o) suggests that bodies, machines, and inscribed marks are all organizations of parts that can be reorganized. The problem of *composition* for such a poetics is thus to ensure that every single "P / art" is a crown. If the "Adam" that we might hear through "Anatom" (and also in Bergvall's title, *Goan Atom*) would seek to claim every part as a subject of his sovereignty, then the task of Bergvall's poetics is to de-subjectify those parts. The problem of politics and aesthetics that Jacques Rancière terms "the distribution of the sensible"[47] is to seek an organization of bodies, whether those of texts or sociopolitical situations, in which the parts of a collective attain their singularity not as a part of a whole to which they are subordinated but as a distribution of singularities that displaces the whole. When "Every single P / art is a crown / to Anatom," each will claim as much sovereignty as any other. The problem of collective articulation is how to organize and subversively disorganize the bounded field of relational elements that collectively constitutes a text, a body, a body politic.

It is because of passages like this that I think "objectism," as an ontology and a poetics, offers a helpful frame through which to grapple with the conceptual and literary-historical stakes of Bergvall's work. And it offers an equally helpful frame in which to take up *Goan Atom*'s engagement with biotechnology. Nanoscale materials science is able to traverse the "dry" and "wet" worlds of solid-state chemistry and molecular biology because the scale at which it operates enables it to address the particulate substratum of these environments, the relational units of which they are composed, as the materials with which fabrication works. "When we work at the nano-scale," writes Eugene Thacker in *Biomedia*, "the accepted differences between living organisms and inert objects become to some extent, irrelevant. What matters is not the macroscale status of an object . . . but rather the molecular characteristics that such objects express at the nanoscale."[48] Richard Doyle refers to this as the "postvital" condition of the body, "a body without life" that has been "recast as an effect of a molecule."[49] At the nanoscale extremity of this postvital approach to molecular biology "all objects," argues Thacker, "including the human body—are simply arrangements of atoms. It then follows that making any object requires a technology with a precision control over atomic arrangement—the capability to structure matter atom by atom."[50]

Nanoscale fabrication is able to incorporate structural models from either biology or solid-state chemistry and to work at their intersection by dealing with these models in terms of their atomic constituents, thus attempting to organize structures from the bottom up. But working from this minimal starting point also poses a problem for these processes of fabrication, since it tends to neglect traits of the collective environment from which these structures derive and in which they will have to function. It makes no sense to address nanoscale matter in terms of secondary qualities like "wet" and "dry," since "bulk properties such as viscosity and friction are not defined for discrete atomic ensembles."[51] But such ensembles nonetheless participate in collective environments that *are* characterized by such properties, and any fabricated structure—whether characterized at the nanoscale or not—will have to be adapted to such participation. It is now relatively easy to map the sequence of amino acids in a protein molecule at atomic-scale resolution using X-ray crystallography. But it is very difficult to predict and control how such a protein molecule will fold into a globular structure due to the hydrophobic effects of exposure to water molecules, like those found within the interior of a cell. And it is precisely the problem of protein folding that bionanotechnology has to account for if it is to create "designer" proteins that actually carry out the tasks for which "natural" biological macromolecules are so well adapted. As Goodsell writes,

> A major hurdle must be crossed before bionanotechnology will have general applicability: We must be able to predict the folded structure of a protein starting only with its chemical sequence. Without this ability, we will merely shadow evolution, poking and prodding existing proteins until they are changed into something that we want.[52]

No doubt because these problems confronting its practical applications are intractably difficult, the rhetorical strategies and conceptual schema attendant upon nanoscale engineering appear calculated to evade the task of directly confronting them, a confrontation that would involve thinking the complex codetermination of the *object* and the *collective*. If the *technics* of nanoscale fabrication enables an indifferent address to any given body as a collective of objects, the *poetics* of objectism addresses the problem of how to operate at this limit of fabrication without giving way to reification and reductionism, to objectification.

FROM PROJECTIVE TO PERFORMATIVE

Like Whitehead, Olson wanted a term to refer to an ontologically univocal unit that is inherently relational, differentiated not *from* other units but rather *through* complex interdependencies: a *relatively* determinate unit that is active, provisionally stable, and saturated with movement and energy even when it seems to be utterly static. He tried to rehabilitate the term "object" for that purpose, approaching poetry as a means of composing "objects in field" in such a way that particular differences among objects are sustained by the relationality of their being in common, and in which the univocity of their physical being is sustained precisely through the articulation of these differences. That is, physical being is both univocal and differentiated, *determinate*, because the relationality of objects in field at once draws them together into a collective "incompletion in process of production"[53] and distinguishes them as singular. But as has been well documented, one of the major problems with this apparently promising program is that Olson's work fails to adequately engage gender and sexuality as specificities that must matter to any thinking or practice of equality and difference, even—or perhaps especially—when it is "the 'body' of us as object"[54] that is at issue.[55] "*Man* is himself an object," Olson writes,[56] and his unwillingness to complicate the gendering of that supposedly universal noun does not merely derive from generic usage; it is symptomatic of a masculinism pervasive in his work. If Olson's objectism attempts to constitute a common world of physical bodies through an egalitarian materialism, it fails in that project when the body of "us" as object has to be articulated in terms of the differential genders and sexualities that his collective pronoun leaves uninterrogated.

This is not to say that Olson missed the problem of gender and sexual difference entirely. As Rachel Blau DuPlessis points out,[57] midcentury poets like Olson, Ginsberg, and Creeley were critical of and resistant to dominant models of male subjectivity in the 1950s (the figures of the organization man, the breadwinner, etc.). But it remains the case, DuPlessis argues, that these poets "implicitly or explicitly rejected the possibility of making a bilateral gender critique."[58] They benefited from destabilizing gender norms that applied to men, but they also benefited by reinforcing gender norms as they applied to women. "To see men investigate and even change some gender ideas," writes DuPlessis,

> without their appreciating that women could want, in parallel ways, to investigate, and even change gender ideas is to feel a lost opportunity, one on which we might still be able to make good. For these critical poetries of fifty years gone understood only part of what needs to be known, knew only part of what needs to be done.[59]

I want to argue that the category of the object in Olson's poetics retains some interest for thinking through what "needs to be known" and what "needs to be done." It troubles the boundary between nature and culture, and it unsettles the self-evidence of organic or biological unity. Insofar as the concept of the object evades the essentialist categories of "man," "woman," or "human," how might an investigation of bodies as *differentially configured collectivities of objects* be mobilized by a feminist or a queer poetics? If "the" object is always relational through and through, and therefore an internally articulated collective of relational objects, how might that articulation be disarticulated and rearticulated? These are some of the questions I want to address through a reading of Bergvall's work.

In *Bodies That Matter*, Judith Butler refers to Marx's consideration of the object in his first thesis on Feuerbach.[60] Marx, Butler notes, "calls for a materialism which can affirm the practical activity that structures and inheres in the object as part of that object's objectivity and materiality."[61] Butler argues that

> according to this new kind of materialism that Marx proposes, the object is not only transformed, but in some significant sense, the object *is* transformative activity itself and, further, its materiality is established through this temporal movement from a prior to a latter state. In other words, the object *materializes* to the extent that it is a site of *temporal transformation*.[62]

The "new kind of materialism" Butler attributes to Marx, for which "the object *is* transformative activity itself," is precisely the kind of materialism we find in Olson's objectism, and these are the terms on which I want to read Bergvall's feminist/queer poetics: as a materialist poetics positing the object as inherently active, temporally mutable, and relational—rather than inert, self-identical, and passive. Butler usually speaks of *bodies*, but here she seems to be asking: What if objects mattered as much as bodies? Or perhaps: What if bodies mattered *as* objects? What sort of gender trouble would that cause? If the opportunity to ask that last question was missed or elided by Olson, it is opened up again by Butler who (like Olson) thinks of what a body *is* in terms of what it *does*. If Butler wants to challenge the biological self-evidence of the body by arguing that gender is a mutable *series of actions*, and that gender is the discursive matrix within which "sex" is produced, then Butler's interest in Marx's understanding of the object as transformative activity is consonant with her understanding of the body. How would the development of gender and sexuality studies have been altered if Butler's landmark book were titled *Objects That Matter*? What if we considered not only Olson's theory of projective verse but also Butler's theory of gender performativity in terms of "objectism"?

If we put those two objectisms together, we might end up with a book something like *Goan Atom*. Theorizing gender as "a complexity whose totality is permanently deferred, never fully what it is at any given juncture of time," Butler calls for a rethinking and a practice of the body as an "open assemblage that permits of multiple convergences and divergences without obedience to a normative telos of definitional closure."[63] Confronting a technoscientific conquest of wet and dry worlds that addresses the body as such an open assemblage, while nonetheless subjecting it to a normative telos, *Goan Atom* mobilizes the fractured breaks and excessive flows of an unfixed body, urging, in its closing lines, an improper labor of performative misarticulation:

> workit baby
>
> and the spac eB ween's solids
> & the spac in solDis peed s Peech
> (*GA* 76)

Part of what "remains to be done" is to understand how such a poetics of (mis) articulation critically engages the contemporary ideological and technoscientific conditions under which normatively gendered bodies are reproduced. Though I hardly want to argue for Olson's "influence" on Bergvall, I do want to suggest that Bergvall's writing takes up and begins to make good on this "lost opportunity" in Olson's poetics. Her writing exposes the limits beyond which Olson's objectism was unwilling or unable to go, and it invents the means of working through those limits.

We can measure the distance between Olson's and Bergvall's objectism by juxtaposing the manner in which they articulate their work in performance.[64] Let's return to the passage from "In Cold Hell, In Thicket" cited in Chapter 2 as an early instance in which Olson grapples with the involvement of objectism with the problem of objectification:

> The question, the fear he raises up himself against
> (against the same each act is proffered, under the eyes
> each fix, the town of the earth over, is managed) is: Who
> am I?
>
> Who am I but by a fix, and another,
> a particle, and the congery of particles carefully picked one by another,
>
> as in this thicket, each
> smallest branch, plant, fern, root
> —roots lie, on the surface, as nerves are laid open—
> must now (the bitterness of the taste of her) be
> isolated, observed, picked over, measured, raised

as though a word, an accuracy were a pincer!

this

is the abstract, this
is the cold doing, this
is the almost impossible

So shall you blame those
who give it up, those who say
it isn't worth the struggle?[65]

"Who am I but by a fix, and another," Olson asks, "a particle, and the congery of particles carefully picked one by another." His concern here is with the condition of being an object that he endorses as an ontology and that his poetics takes up as a first principle. The poem is about the fact that there is no getting out of this condition, so the task that Olson's poetics proposes is to go further into it. The problem for Olson is how to save the *object* from *objectification*: how to prevent the reduction of the object to an abstraction by engaging its interior activity and by situating it within the relational field by which it is constituted. He does this by fusing the rhythm of the body with that of the poetic line, separating and distinguishing each breath unit from the others, while holding them together in a co-constitutive corporeal field. The struggle to which Olson refers—the difficulty of preventing "a word, an accuracy" from becoming "a pincer"—is, first, how to single something out (how to relate to it as a particular thing) without robbing it of its singularity; and, second, how to recognize the body as a collective of such singularities—a "congery of particles" or an "arrangement of atoms"—without losing a sense of its integrity as a (relatively) discrete unit.

In the following passage from *Goan Atom*, Bergvall is concerned with similar problems:

What of it
bodymass is heavily funded
Swirling aHeads
roll out of place
mount alterity
part on display
round up
to as perfect a square as Octopus
ever canned
be
roll on roll on

MOTion
begs out of
g
GA
g
ging
Dis
g
orging
b
loo*
p
uke
s
Uck
ack
ock
S
OG
ex
Creme
ental
eaT
ing sp
Am
mon
Am
mon
sp
you
d out
the 1 called
the one called
wholly
quartered
Beloved
Beloved
chok
en the Egg
SP

in
your
arm
to ram
my Hoop
of larm
b
(click)
Look
!
You yo
Footy
facey
(click)
yoyo
and a leg loops
into the
BAC of the
my thRoat
and Ro
und and Ro
und and the Rolling Eye up we gl gloue
in the bl**dy dans hole (*GA* 47–49)

"Bodymass," as Bergvall puts it in the second line of this passage, is "heavily funded." The body is so quantified and calculated by capitalist subsumption that alterity, she implies three lines later, might seem little more than a part to be played or to be mounted and put on display ("mount alterity / part on display)." Olson recognizes the *problem* of objectism in the content of his lines, while attempting to *solve* that problem through the breath poetics that is integral to their form. But Bergvall's disintegrative lines and the disruptive glitches in their articulation index the breakdown of the formal coherence that Olson's breath poetics attempts to salvage.

If Olson's objectism is *projective*, and Bök's objectism is *subtractive*, Bergvall's objectism is *performative*. Bergvall performs the condition that Olson describes: the decomposition of the body into a congery of particles, registered here as graphemic and morphemic fragments that remain stubbornly resistant to any reintegration. While the relational interface situating the body within the larger field of objects that Olson attempts to establish depends upon respiratory rhythms, the rhythms of flows and blockages that are of interest to Bergvall pertain to menstruation, excretion, conception, and parturition. But these rhythms are not sufficient to establish any sense of embodied plenitude or formal coherence. Bergvall's infantilizing idioms toward the end of the passage suggest that at the very moment of its birth, the body is already decomposed or photographically framed into objectified components, or body *parts*. "Footy / facey / click," she writes. These are the parts that seem to be lodged in the back of the poet's throat, blocking the respiratory rhythms upon which Olson's formal recuperation of an integral field relies. The struggle, for Olson, is not to save the human or the organism from the condition of the object but rather to persist within a condition of objectism without thereby giving way to objectification. But if Bergvall finds that Olson's solution to that problem—breath poetics—is blocked in her case, blocked by the irresolvable *decomposition* of the body, then the question becomes how to work through that blockage rather than pretending it isn't there. *Goan Atom* thus proceeds within a condition of textual atomization. This condition of fragmentation, however, also lends itself to a project of collective reorganization, unsettling the fixity of the body, and Bergvall thus mobilizes it as a method of poetic fabrication that might be used to rearticulate the problem that Olson poses: that of "the 'body' of us as object."

ENTER **DOLLY**

Whereas Olson's "us" is purportedly generic but implicitly masculine, the feminist stakes of Bergvall's rearticulation of the collective body are most immediately obvious in her engagement with the figure of the doll. The subtitle of *Goan Atom* is "Doll," or, in an earlier version, "jets-poupee," which, translated literally, means "doll spurt." An epigraph to the book informs us that "Anybod's body's a dollmine" (*GA* 5). If we use the volume as a flipbook, fanning its pages in reverse order, stray boldface letters printed along the extreme left hand margin of the verso leaves spell out the phrase "dolls should be seen" (a quotation from Gertrude Stein's *How to Write*). And a character named **DOLLY**, the cloned sheep named after Dolly Parton, appears as one of several dramatis personae populating the pages of the book as if it were a contemporary incarnation of commedia dell'arte (**DOLLY**'s costars include **HEADSTURGEONS**, **FISHMONGRELS**, **A CO CALLED MOO**, the **EVERY HOST**, and a **GROUP OF CORPOREALS**). Bergvall has explained that her interest in the figure of the doll emerged from an examination of the "articulated dolls" built and photographed by the Surrealist artist Hans Bellmer in the 1930s. The first of Bellmer's articulated dolls, produced in Berlin in 1933, consisted of a molded torso made of flax fiber, glue, and plaster, with a masklike head and a wig of long hair covered by a beret. The midsection of the torso is cut away, revealing a mechanical system of gears and levers, and the right leg of the doll appears to have been "amputated" and replaced with a prosthetic limb. In Bellmer's 1934 photograph of this figure, the doll is posed in front of a dual-perspective anatomical drawing, while a double exposure of Bellmer himself leans broodingly over his composition. Another photograph shows the disassembled anatomical components of the doll neatly arranged as if for inventory, and Bellmer would continue throughout the 1930s to produce and photograph dozens of dolls in various arrangements (Figs. 51 and 52).[66]

In an interview conducted in 1999, while she was writing *Goan Atom*, Bergvall comments on the relation of her work to Bellmer's:

> [In Bellmer's work] the whole certainty of the female body, the female gender (because he used a "girl" doll) becomes problematized. Even though his take remained very misogynistic and even paedophilic, the whole notion of the fixity or the stability of the body does begin to break down. . . . The Doll project for me was a way of playing with language, of disarticulating language at the level of the syllable very often. It was also a way of setting up word games, puns—some of them fairly bad, others very sexual, erotic, of adding on games where you suddenly switch into French. This is a way of thinking about this multiple body. I suppose, this unfixed body that for me today, at the end of the nineties rather than the surrealist thirties, has my own take on it. This has a lot to do with issues of gender but also to do with issues of genetic engineering . . . with the links that are being made in our collective imagination about gender and sexuality at the moment which Bellmer wasn't able to tap into in the same way.[67]

For Bergvall, Bellmer's dolls anticipate the "unfixed" and "postvital" condition of the body engendered by contemporary biotechnology, while also suggesting a poetic means of addressing that condition through the "disarticulation" of linguistic units. For Bergvall, after Bellmer, "the whole certainty" of the body *as* whole, or of grammar as coherent order of articulation, gives way to a multiplicity that is open to intervention and reorganization.

Thus, when **DOLLY** enters *Goan Atom*, she enters entered:

> Enter **DOLLY**
> Entered enters
> Enters entered
> Enter entre
> en train en trail

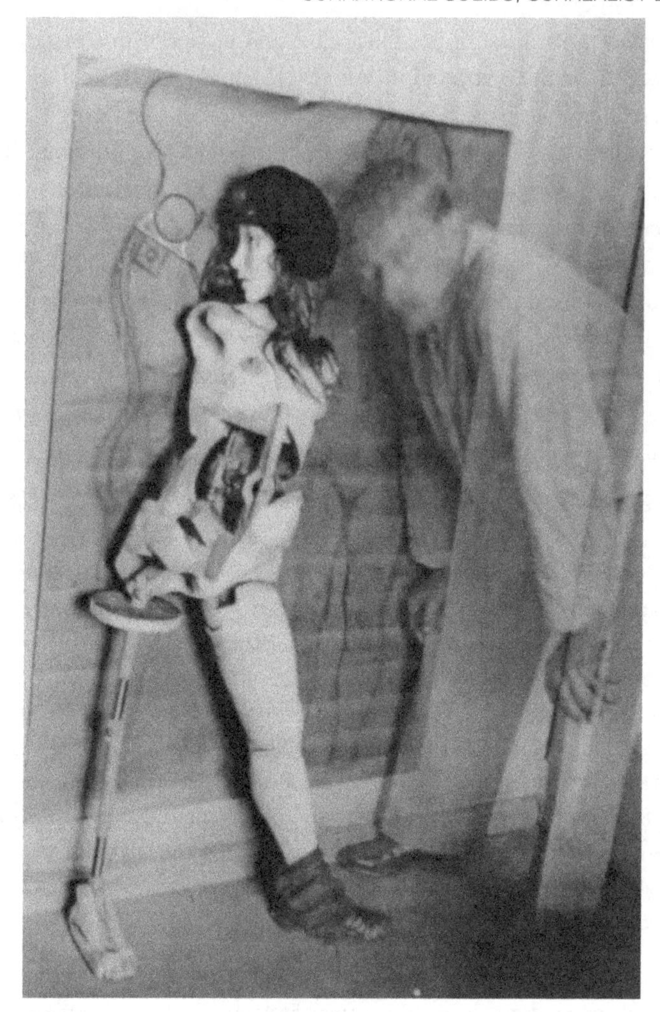

Figure 51 Hans Bellmer, *Untitled*, 1934. Black-and-white photograph. Courtesy of Art Resource and SODRAC.

en trav Ail Aïe
La bour La bour La bour
wears god on a strap
shares mickey with all your friends (*GA* 23)

If the mise-en-scène of pastoral poetry evokes a rural lad tending his sheep as he composes verses, Bergvall updates this scene for the twenty-first century, in which it is the sheep itself that is composed in a laboratory. Bellmer's doll,

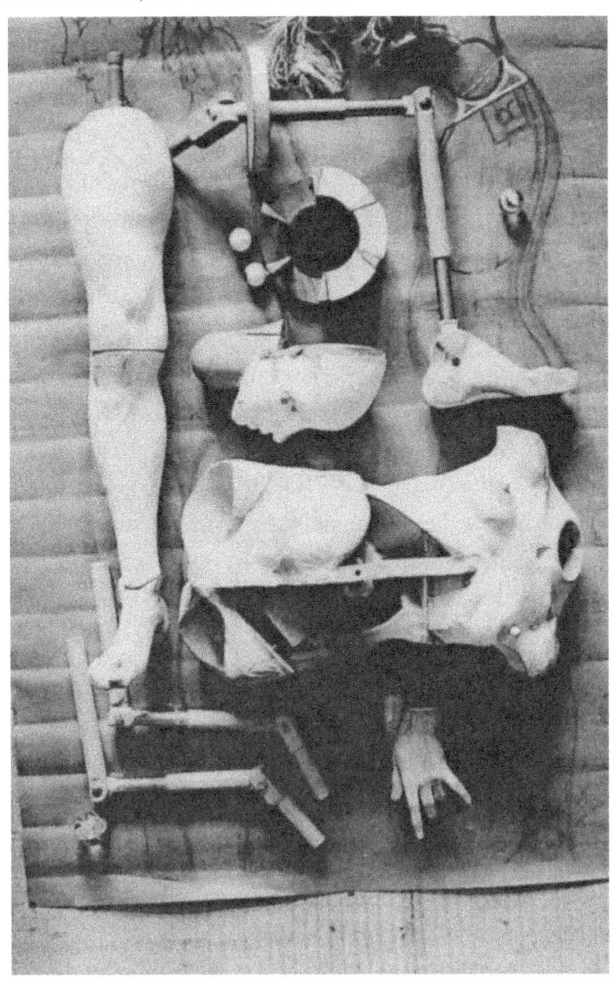

Figure 52 Hans Bellmer, *Untitled*, 1934. Black-and-white photograph. Courtesy of Art Resource and SODRAC.

reborn as the darling of contemporary technoscience, cannot enter that scene without *already* having been entered by the machinations of genetic engineering that brought her into the world. And as Bergvall's punning fusions of French and English make clear, as soon as Dolly enters the world "she" is already working. "In the biotech industry," argues Eugene Thacker in *The Global Genome*, "labor power is cellular, enzymatic, and genetic." "Not only does the biotech industry adopt an instrumental approach to biology," Thacker explains, "but in doing so it also locates the productive life activity of biology at the level of cellular metabolism, gene expression, protein synthesis, and so on . . . it is biological and economic, a biomaterial labor power."[68] When Dolly enters, "Enter, entre" gives way to "en train en trail," situating spirit or drive (*entrain* in French) within the viscera of the body (or *entrail* in English). *Entrain* is a term used in the French expression "travail avec entrain," to work energetically, or with spirit. But here Bergvall points up the embodied *duress* of biomaterial labor power by following particles of the word *travail* with the exclamation "Aïe"—roughly the French equivalent of the English "ouch." The separation of "La" from "bour," exposing the feminine article in "labour," exposes childbirth as reproductive labor. This is a move that is equally pertinent to second-wave feminist struggles over wages for housework or to the contemporary domination of women's time and bodies in the sweatshops of Latin America or Southeast Asia. La Bour is also the brand name of a self-priming pump used in the mining industry, featuring a special mechanism to remove potentially disruptive "entrained air" from the water that circulates through it. So the term "La Bour" figures labor as "feminine" praxis or capacity, as the exploitation of women's bodies, and as a kind of perpetual motion machine designed to eliminate any potentially disruptive elements that might disturb its operation.

What is at issue here is the imbrication of sex and gender, body parts and parts of speech, in the means of production and of reproduction. Discussing chromosomal definitions of "sex" in *The Epistemology of the Closet*, Eve Kosofsky Sedgwick makes the basic point that perhaps "*the* primary issue in gender differentiation and gender struggle is the question of who is to have control of women's (biologically) distinctive reproductive capability."[69] The figure of Dolly indexes the technocultural moment at which the capacity to produce and to reproduce "female" bodies is claimed by men, even as those bodies are put to work for the production and reproduction of cultural and monetary capital both in and outside of the laboratory. Thus labor itself "wears god on a strap" insofar as the capacity to reproduce is claimed as phallic, even as that phallic capacity is exposed as artificial. The phallus operates here as a token of, supplement to, or substitution for biological sex organs, making those organs obsolete even as their gendered division is reproduced and sustained as *part and parcel* of the status quo:

AH YES
puts in the **EVERY HOST**
but sheeped
like a dolly

part out part ed
partout prenante
every little which way
through the mid-
Come 'n
gain a bit
Come a kiss
(is made of this)
: it's a girl
Come a kiss
: and it's not
In fact it was
inconvenient (*GA* 54)

Dolly is "sheeped *like* a dolly," like a toy that reinforces heteronormative divisions of sex and gender based on the determination that "it's a girl." Bergvall plays upon Dolly's phantom surname, Parton ("part out part ed"), to suggest that the reciprocal determination of sex and gender, supposedly a matter of physical body parts, is in fact a matter of distribution. Gender is a matter of how the "parts" of the body are seized upon, parceled out, and put together. The partitioning of a body is "partout prenante," completely or entirely *gripping*, insofar as what any body "is" is supposedly determined by whatever one finds or doesn't find "through the mid-." Bergvall's work is *feminist*, insofar as it critiques the biological determination of the "female" by particular body parts or by singular genes, abstracted from the embodied contexts in which they operate. Her work is *queer* insofar it engages the gendered body and the material text as collectivities that, because they are articulated into differential parts, can be rearticulated in a multiplicity of ways that are irreducible to binary determinations of sexuality or the normative rules of grammar. The "inconvenience" mentioned at the end of the stanza above might be taken to refer to the irreducibility of corporeality to binary genders and to the excess of sexuality over discrete sexes or sexual orientations.[70]

One can imagine a reading of Bergvall's work that might proceed along the lines of Julia Kristeva's *Revolution in Poetic Language*.[71] One might say: Bergvall's disassembly of normative grammar and monolingual propriety, her liberation of the grapheme and the phoneme from their subordination to the referential semantics of the word, reasserts the rights of the semiotic against the domination of the symbolic order. Bergvall's writing, such an approach might proceed, releases and indexes the pulsations of pre-Oedipal libidinal flows characteristic of pre-genital sexuality and prior to the disciplining of polymorphous orality into articulate speech. All of this may very well be the case, and certainly Kristeva's work is an important influence on Bergvall's poetics. But the contexts in which Bergvall situates her formal strategies—her engagement both with Bellmer's dolls and with biotechnology—make it hard to see those formal strategies as a celebratory reassertion of sublimated pre-Oedipal *jouissance*. The fracturing of Bergvall's language, the recombinant operations of cutting and splicing and through which *Goan Atom* disarticulates and rearticulates body parts and parts of speech into novel arrangements: these are precisely the operations of the masculinist projects that her text takes on. Moreover, genetics and biotechnology expose the fact that code, the Symbolic, is of and in the body through and through, that an engagement with the parameters of embodiment within the discourse networks of contemporary technoscience will have to work *through* a symbolic order that cannot be disengaged from any prior real onto which it is inscribed.[72] Donna Haraway has pointed out that gene fetishism, the fixation upon genes or "the genome" as singular determinants of biological traits, operates through the double move of abstracting a master molecule or its function from the larger context in which those functions are performed *and* by abstracting the systemic unity of a total "code" from the corporeal interactions of multiply particular, temporally involved agents. The corporeal urgencies of Bergvall's language remind us that codes are not imposed upon but *of and in* the body that instantiates them, while the combinatory permutations of her poetics suggest that no single corporeal or grammatical unit can certify the unity of a body, its gender, or its sex.

AMBIENT FISH

The link between linguistic and corporeal registers of coded articulation is perhaps most readily evident in the title *Goan Atom*, which offers a preview of the uneasy thematic conjunctions and the recombinant formal strategies that are featured throughout Bergvall's volume. The phonetically discernible project of *goin' at'em* prepares us for the aggressively oppositional attitude characterizing Bergvall's punk sensibility, while the fractured and truncated morphemic remnants of the word *anatomy* suggest the atomized body that the text will both dissect and reassemble (Fig. 53).

The cover art for the book—a stylized green icon resembling an artificial breast, dubbed "Green Nip" in the volume's front matter—reinforces that suggestion, indicating that the fragmentation and objectification of women's anatomies

Figure 53 Caroline Bergvall, "Green Nip." Cover art for *Goan Atom*. Courtesy of Caroline Bergvall.

will be at issue, while the title's phonetic approximation of the ur-masculine name "Adam" implies that the target of Bergvall's impending attack will be the neo-Edenic presumption of any agent posing as the arbiter of proper names and their determinate referents. As Bergvall is surely aware, the first three letters of her title, G.O.A., function as an acronym for Gene Ontology Annotation, an international formal language used by scientists to codify and communicate protein functions for thousands of different species. GOA's searchable online database informs its users that

> one common problem with publications on protein function is that the language used by scientists is not very precise—words often have several meanings, such as "cell". It could mean a "prison cell", a "battery cell" or a "living cell." As anyone who has tried to explain something complicated by e-mail will know this can cause all sorts of confusion. . . . What is needed is a list of words that have only one "official" meaning. This is the purpose of a "controlled vocabulary." . . . [Using a controlled vocabulary] means that scientists will be able to make use of more of other researchers' work with less effort and that means they can do their science quicker and better, which will be a benefit to us all.[73]

Goan Atom's project of *goin' at'em* is directed at precisely this sort of discourse, a "controlled vocabulary" in which discrete corporeal functions are sutured to "official meanings" in the name of communicative expediency and the presumptive universality of scientific progress. But rather than tapping into a condition that is *prior* to the entrance of the body and of orality into the symbolic order, Bergvall's strategy is to work *within* the articulation of the body and the text by rules, practices, discourses, and technologies. Performatively mimicking and exacerbating the division of bodies into body parts, and of orality into parts of speech, her work attempts to retain the plurality of those divisions and discrepant parts while redistributing them into "inconvenient" arrangements. If "objectism" is a useful point of departure from which to approach this sort of practice, that is because it focuses our attention upon relational but discrete units of articulation and upon the difficult project of retaining the multiplicity of those units against their integration into encompassing totalities, controlled vocabularies, or coherent subjectivities.

The performative character of Bergvall's objectism, the formal strategies of mimicry and combinatoric dis-/re-articulation through which she debunks the promissory notes of contemporary technoscience, are perhaps most immediately and aggressively manifest in her piece *Ambient Fish*, which exists in two different versions: as a text piece included in *Goan Atom* and as a flash animation on the Web. *Ambient Fish* might be read as the definitive commentary on the famous "nanobots" of K. Eric Drexler's fancy and the anticipated virtues of their biotechnical applications. These nanobots, Drexler wrote in 1989, will soon be "searching out and destroying viruses and cancer cells," "enter[ing] living cells to edit out viral DNA sequences and repair molecular damage." They will "enable the clean, rapid production of an abundance of material goods." "Nanotechnology," enthuses Drexler, "will give us this control, bringing with it possibilities for health, wealth, and capabilities beyond past imaginings."[74] Frankly, this is the kind of money shot that Bergvall has seen before:

Ambient fish fuckflowers bloom in your mouth   will choke your troubles away
Ambient fish fuckflowers bloom in your mouth   will choke your troubles away
Ambient fish fuckflowers bloom in your mouth   will choke your troubles away
Ambient fish fuckflowers bloom in your mouth   will shock your double away
Ambient fish fuckflowers loom in your mouth   will soak your dwelling away
Alien fish fuck fodder loose in your ouch   suck rubble along the way
Alien fish fuck fodder loose in your ouch   suck rubble a long way
Alien fuck fish fad goose in your bouch   suck your oubli away
Alien phoque fresh fat ease in your touche   watch a getting away
Alien seal fresh pad easing your touch   take the gamble away
To fish your face in the door
a door a door
fuckflowers bloom in your mouth   will choke your troubles away

Ambient fish fuckflowers bloom in your mouth   will choke your troubles away
Ambient fish fuckflowers bloom in your mouth   will choke your troubles away
Ambient fish fuckflowers bloom in your mouth   will choke your troubles away
Ambient fish fuckflowers bloom in your mouth   will shock your double away
Ambient fish fuckflowers loom in your mouth   will soak your dwelling away
Alien fish fuck fodder loose in your ouch   suck rubble along the way
Alien fish fuck fodder loose in your ouch   suck rubble a long way
Alien fuck fish fad goose in your bouch   watch a getting away
Alien phoque fresh fat ease in your touche   watch a ramble away
Alien seal fresh pad easing your touch   take the gamble away
To fish your face in the door
ador ador

fuckflowers bloom in your mouth   choke your troubles away

(*GA* 72–73)

The simultaneously solicitous and threatening refrain "ambient fish fuckflowers bloom in your mouth | will choke your troubles away" undergoes a series of semantic and phonetic displacements, from "Ambient fish fuck flowers loom in your mouth | will soak your dwelling away" to "Alien fuck fish fad goose in your bouch | suck your oubli away" to "Alien seal fresh pad easing your touch | take the gamble away." In performance, Bergvall reads the piece in a disconcertingly calm, smooth voice that works in tension with the profanity of its vocabulary.[75] The form of the piece, the iterative permutations to which it subjects its material, analogically deploys the recombinant cutting and splicing operations of

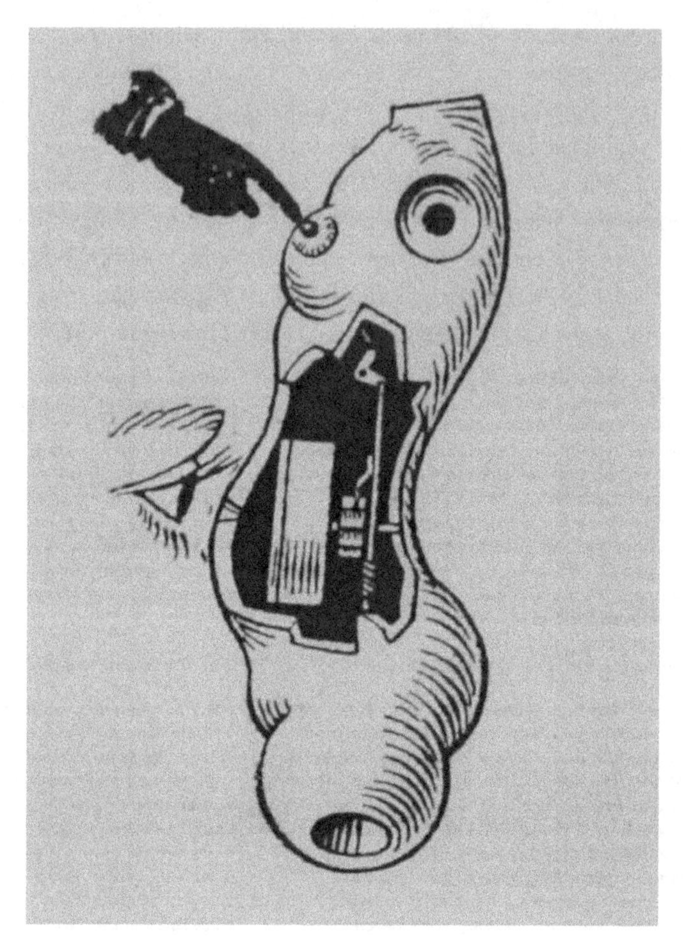

Figure 54 Hans Bellmer, *Untitled*, 1934. Linocut on pink paper. Courtesy of Art Institute of Chicago.

genetic engineering. The content of the piece, its ominous evocation of ambient aquatic invaders breaching the interiority of the body through its orifices, links such speculative visions as Drexler's (along with the relatively mundane molecular probes, microcatheters, and drug delivery systems of twenty-first-century biomedical technology) to the language of advertising and the omnipresence of pornography in the online society of the spectacle.

The electronic version of *Ambient Fish* confronts that online context on its own turf: the Worldwide Web. The visual framing of this version engages with one of Hans Bellmer's most famous images, in which an iconic hand points to

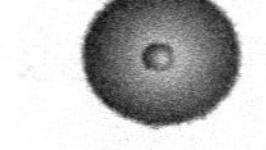

Figure 55 Screenshot from Caroline Bergvall, *Ambient Fish*, 1999. Courtesy of Caroline Bergvall.

or presses the nipple of a female torso like a button, while a disembodied eye stares through the navel of the torso at a mysterious interior mechanism (Fig. 54). Early in *Goan Atom*, Bergvall quotes a passage from Bellmer's essay "Memories of the Doll Theme," in which he instructs himself to "lay bare suppressed girlish thoughts, ideally through the navel, visible as a colorful panorama electrically illuminated deep in the belly."[76] On the introductory web page of the electronic version of *Ambient Fish*, we find two vertically aligned red buttons. If we click on the top button, the same stylized artificial breast icon that we find on the cover of *Goan Atom* appears and rotates under the finger of the cursor (Fig. 55). When we click on the lower of the two buttons, we gain access to the "panorama electrically illuminated deep in the belly" that Bellmer wanted to "lay bare," as Bergvall imagines that panorama might be illuminated in 1999 (Fig. 56).

In the Flash version of *Ambient Fish*, the female body is reduced to a grid of reified, cathected body parts that are exchanged, one by one, for the fragmented utterances of an alternately articulate and disarticulated voice. Detached breast for detached voice: one *objet a* for another. Bergvall's grid of generic partial objects plays up the Object Oriented Programming of the code in which her piece was written, fusing the base of informatic production with the user-friendly superstructure of our icon-driven operating systems. The poem grapples with a technoscientific context in which the functions of different bodies are indifferently reduced to a click of the mouse and in which the body itself "becomes what we might call 'programmable matter,' a materiality characterized by a constructionist logic, and a highly discrete, combinatory mutability induced through the intersection of molecular biology and mechanical engineering."[77]

Certainly *Ambient Fish* engages in an elaborate critique of objectification. But even as Bergvall's piece registers the reduction of the body to a collection of *abstract* objects, it simultaneously deploys that reduction as the very means of its *concrete* poetry. Bergvall seems always to be equally concerned with the objecthood of the body and of body parts as something that happens *to* bodies, that *makes* them "female," and as a phenomenon through which what Bergvall calls the multiple or unfixed body might be rearticulated *for* and *by* a feminist and queer poetics. In this latter case, objectism would function not as a way out of objectification but as a way of engaging the ramifications of the latter and of reconfiguring the decomposition of the integral body in which it results. That is the project through which Bergvall performs a *détournement* of Bellmer's articulated dolls, and in doing so she also succeeds in reinventing an objectist "stance toward reality" for a twenty-first-century feminist and queer poetics.

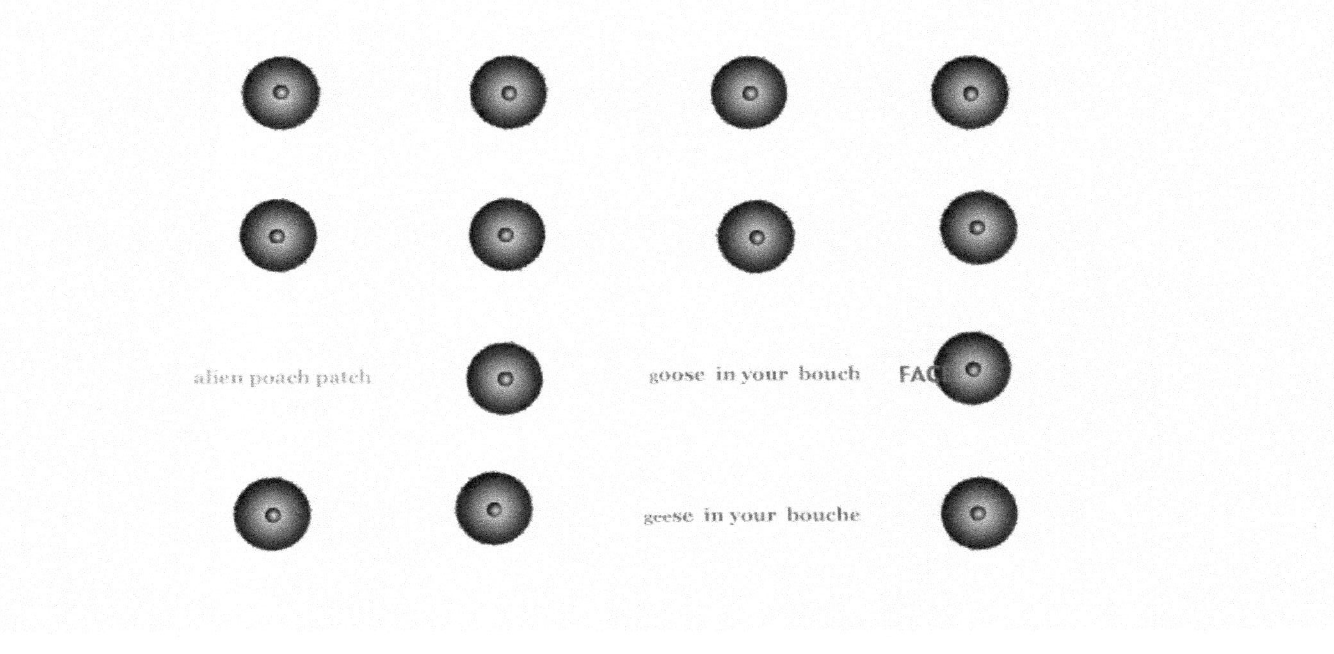

Figure 56 Screenshot from Caroline Bergvall, *Ambient Fish*, 1999. Courtesy of Caroline Bergvall.

5

THE SCALE OF A WOUND: NANOTECHNOLOGY AND THE POETICS OF REAL ABSTRACTION IN SHANXING WANG'S *MAD SCIENCE IN IMPERIAL CITY*

Go in fear of abstractions.

—Ezra Pound, "A Retrospect"

Words arrive in triangles and squares and leave in circles without any lasting registration in the brain. However my wounds just stay. They never came. I never let them go. My abstract wounds with no definite shapes and locations.

—Shanxing Wang, *Mad Science in Imperial City*

WORK NANO, THINK COSMOLOGIC

Midway through Shanxing Wang's 2005 volume *Mad Science in Imperial City*, we encounter the following imperative: "work nano, think cosmologic."[1] What does this mean? What are the implications of this imperative for poetics? Unfolding a reading of *Mad Science* from these questions, I want to show that what is ultimately at issue in the imperative to "work nano, think cosmologic" is a daunting problem: how to take the measure of a wound. Wang's book makes clear that the wound in question not only belongs to a person or a place; the wound of which his work takes the measure inheres in history.

Wang was born in the Shanxi province of China in 1965, "on the eve of the Cultural Revolution."[2] He studied mechanical engineering at Xi'an Jiaotong University in the 1980s. During that decade, he became active in antistate political struggles in Beijing and participated in the mass demonstrations at Tiananmen Square in 1989. In 1991 he moved to the United States to pursue a Ph.D. at Berkeley, following which he joined the faculty of mechanical engineering at Rutgers, where he became engaged with nanoscale fabrication around 2000. In 2001 he began taking creative writing classes on campus, and eventually he gave up his professorship at Rutgers in order to focus on writing, taking further workshops at the St. Mark's Poetry Project in New York and Naropa University in Colorado. The outcome of this radical decision would be *Mad Science in Imperial City*, which won the Asian American Literary Award for poetry in 2006.

Mad Science is an extraordinary text, "a brilliant gem dropped through the keyhole from an alternate universe," as Brian Kim Stefans describes it.[3] At once a memoir, a recessed historical epic, an awkward love story, a 'pataphysical experiment, it draws upon the resources of Wang's scientific knowledge to grapple with the traumatic conclusion of his political commitment in the 1980s and his experience of parallels between totalitarian China and the political climate of the United States following September 11, 2001. Constellated by equations and scientific diagrams, *Mad Science* attempts a mathematical formalization of historical processes and an algebraic encoding of subjective experience. The effort of Wang's speaker is to inscribe the affective and political terrain of a compound city that seems to be everywhere and nowhere, an "imperial city" sprawling between the two discrepant upheavals the East and West constituted by the Tiananmen Square massacre of June 4, 1989, and the attack on the World Trade Center of September 11, 2001. Wang writes memoir as though a mimesis adequate to the contingent encounters and emotional investments he tries to record requires the formal language of the sciences: "I turn the pages of my math books to look for him, solve for him" (*MS* 19). This is a method as productive of formidable defamiliarizations as it is of beautiful prose and heartbreaking simplicities:

> I write winter as he says October yes it's the 1st day of October but the interval bounded by June and winter was null after that because the implosion in the heart of the capital through fission of heavy elements of uranium-like emotions created the instantaneous nuclear winter inside the head by blasting through the silenced channel of the sore blown deep throat a cloud of soot and dust of the disoriented disintegrated feelings which blanketed the tiny planet of the head one layer after another from absorbing any ray of sunlight of revelation and rationalization and reduced its interior temperature of apprehension and expression to -273°C in both its half-spherical chambers. (*MS* 128)

From June to October 1989: this is a description of summer—a summer between the spring of Tiananmen and the autumn of its aftermath. It is a description of the seasonal affective disorder induced by the "heavy elements" of political repression, the "uranium-like emotions" of a broken revolution, the "nuclear winter" of lost comrades, the null set of the collective, the absolute zero of history. The problem of Wang's text is how to craft a historiography adequate to this kind of season, the experience of which is at once immediate and delayed, as overwhelmingly direct as it is ungraspably oblique.

The inextricability of Wang's scientific knowledge from his writing means that his poetry turns not only upon a hinge joining the twentieth and twenty-first centuries, East and West, totalitarianism and "democracy," politics and love, science and poetry, but also *working* and *thinking*: the nanoscale fabrication projects he was pursuing at the end of his career as an engineer and the speculative, cosmological scope of his concerns in *Mad Science*. When asked about the role of nanotechnology in both his engineering work and his poetry in a 2006 interview, Wang responded as follows:

> I worked at the macro scale, working with industrial machines and automobiles, at the level of the everyday macro world. And then we ventured into laser manufacturing at the micro level, and then finally began working at the nanoscale. But ultimately manufacturing, if you trace it back all the way, if you regress all the way back, it begins with the Big Bang. That's the first "manufacturing" really—you know, photons, random fluctuations of temperature, the formation of elements. Hydrogen forms, and titanium, the other heavy elements. Gravitation, planets, stars. That's where we *really* come from. And we can try to trace ourselves back to that source of our being. Not just where we are from spatially, in terms of location.[4]

For Wang, this is what it means to "work nano, think cosmologic": to think through the significance of nanoscale fabrication—and of manufacturing in general—from a perspective vast enough to include the origin of the universe. "I am always concerned with history," he adds.[5]

But "work nano, think cosmologic" is also an implicitly parodic formulation, heavy with bathos. Considered as a general imperative to work and think in a manner accounting for all orders of scale, it registers genuine concerns running throughout Wang's book, with which his whole life is bound up. As a phrase, however, it sounds like an advertising tag line, a specimen of the sort of technobabble that might serve, for example, as a slogan for the National Nanotechnology Initiative (NNI), to which the U.S. government has devoted billions of dollars since the turn of the millennium. Initiated in 2000 by Bill Clinton, the NNI was advertised at its inception by a brochure bearing the unintentionally comic and curiously retrograde image of an atomic landscape receding toward a

Figure 57 "Shaping the World Atom by Atom." Promotional brochure for the National Nanotechnology Initiative. Courtesy of the National Nanotechnology Initiative.

cosmic horizon, accompanied by the slogan "Shaping the World Atom by Atom" (Fig. 57).

Wang remarks that the inception of the NNI marks the moment at which "nanotechnology became part of our national character."[6] That this national character is inherently *imperialist*, part of the compound "imperial city" of Wang's title, was made clear by the backdrop of Clinton's speech at Caltech, inaugurating the NNI. Below the heading "Investing in Science & Technology for a Strong America," the image showed a map of the western hemisphere constructed with individually positioned gold particles. In this image of nanotech as the new frontier of technoscientific progress and economic growth, the spoils of the old New World figure nanoscale matter as a new New World, one that remains available for discovery and colonization despite late capitalism's exhaustive exploitation of natural resources. The race of competing nation-states to make good on nanotechnology will be the new Gold Rush; "Shaping the World Atom by Atom" becomes the millennial manifestation of manifest destiny. Considered alongside the atomic IBM logo of 1989, which made clear that there is nothing so small as to evade the dominance of International Business Machines, these images figure nanotechnology as the organon of imperial conquest at all orders of scale. Given that the slogan of the IBM corporation is "THINK," the phrase "work nano, think cosmologic" might be taken to unpack IBM's exercise in nanoscale corporate branding as the implicit ideology of the NNI's cosmological and geopolitical imagery. Wang's formulation, then, not only indexes a subjective problem concerning the relation of laboratory life ("work nano") to intellectual ambition ("think cosmologic"), a genuine crux of his own biography. It also correlates that subjective tension to economic and political determinations concerning corporations and nation-states, determinations within which an important episode in the history of technoscience was embedded at the turn of the millennium.

With this context in mind, we can note that the phrase "work nano, think cosmologic" transposes an earlier imperative, "Think Globally, Act Locally." Popularized by Buckminster Fuller and the *Whole Earth Catalog* in the 1960s,[7] the latter slogan might be taken to represent a mid-twentieth-century surge of ecological consciousness attending the growth of information networks and the incipient political-economic conditions of globalization: the moment not only of the *Whole Earth Catalog* and Marshall McCluhan's "global village" but also of Robert McNamara's application of systems analysis to corporate organization and international policy, as well as the rise of the International Monetary Fund and the World Bank in the wake of the Bretton Woods Conference. That is, the ecological imperative to condition local action in accordance with global perspectives is deeply intertwined with the emergence of information theory, systems theory, game theory, and the complicity of these with the role played by militarism and finance in the rise of the United States as global hegemon. Reinhold Martin has diagnosed this polyvalent mid-twentieth-century vogue for systems thinking as "the organizational complex": "the discursive formation from which both the technomilitarism of control systems *and* proposed antidotes to this militarism . . . sprang during the 1950s and early 1960s."[8] As one of these proposed antidotes, the slogan "Think Globally, Act Locally" emerges from and feeds back into the organization complex of the postwar world order, determined as it is by the interconnected political, economic, and ecological conditions of late capitalism.

It is worth noting that the development of materials science in the twentieth century provides an apt illustration of relations between the local and the global characteristic of the postwar world order. As Ernest Mandel points out in *Late Capitalism*, the rising cost of producing raw materials in underdeveloped countries during the first half of the twentieth century led to increased production of *synthetic* materials in metropolitan centers (synthetic rubber, synthetic fibers).[9] From 1938 to 1965, the share of production of synthetic fibers in the world production of textiles rose from 9.5 to 27.6 percent, while the share of the production of synthetic rubber in the total world production of natural and synthetic rubber rose from 6.4 to 56 percent.[10] This increased production of synthetic materials in industrialized countries contributed to a relative decline in the demand for natural raw materials extracted from the Third World. Thus the *local* production of synthetic raw materials in the First World contributed to the opening of Third World labor markets for the production of finished goods rather than the extraction of raw materials—a transition that helped to enable increased industrialization of the Third World ("development") during the postwar economic boom and the offshoring of manufacturing labor. Such structural modifications of world labor markets illustrate the systemic interconnection of local and global production, correlated both to technological affordances and to the demands of capital accumulation. As Mandel points out, the entire global division of labor created in the nineteenth century—in terms of the relation between First World industrial production and Third World materials extraction—was transformed in the middle of the twentieth century. One enabling condition of this transformation was the technological capacity to fabricate synthetic materials in metropolitan centers.

One might argue that Wang's transposition of "Think Globally, Act Locally" into "work nano, think cosmologic" shifts the problem of cognitive mapping the former phrase indexes from the register of geography to that of scale.[11] "Work nano" *shrinks* the "local" site of production to molecular and submolecular scales at which familiar physical laws of macroscale matter (let alone the human perceptual apparatus) do not operate. "Think cosmologic" *expands* the contextual frame of the "global" to orders of magnitude at which efforts to theorize a political-economic totality give way onto scientific investigations of dark matter, black holes, and multiple universes. These shifts pose rather different problems of cognitive mapping. Barrett Watten has retrospectively argued that the problem of the referent taken up by Language Writing in the 1970s—the problem of writing in such a way that the "radical particularity" of syntax and form resists referential transparency—responded to displacements of local production by the incipient conditions of globalization.[12] With reference to Watten's argument, Joshua Clover remarks in an interview that "language writing helps us rethink what might be revealed and concealed by the phrase 'act locally'—but, and this is my concern, at the expense of truly thinking globally. That's another way to phrase the demand on contemporary poetry, rather than simply posing it as a matter of aesthetic development."[13] For Clover, "the demand on contemporary poetry" is to not retreat from the level of totality ("thinking globally") in order to secure the local action of "radical particularity." Should we understand Wang's imperative as a formulation of "the demand on contemporary poetry" to address the problem of cognitive mapping not only in terms of "global" totality or "local" particularity but also in terms of atomic and cosmological orders of scale? If so, what sort of problem of the referent might such a demand entail?

INTELLECTUAL AND MANUAL LABOR

To begin grappling with this question, we need to recall that Wang's formulation links not only discrepant orders of scale (the "nano" and the "cosmologic") but also the activities of working and thinking. Thus the phrase not only registers the somewhat cloying imperative of a technoscientific zeitgeist, it also denotes a relation between work and thought, intellectual and manual labor, within the post-Fordist regime of accumulation by which Wang's scientific and poetic practices are conditioned. The flood of venture capital attending the late 1990s tech bubble, for example, helped generate the surge of interest in nanotechnology during which Wang's engineering work shifted to nanoscale fabrication, which then comes to play a major role in his poetry.[14] If the promotional flyer for the NNI figures the injunction to "work nano, think cosmologic" as a matter of "shaping the world atom by atom," and therefore as a matter of *manual* labor, we should also bear in mind the conditions of *intellectual* labor necessary to propel that project. Indeed, *fabrication*, whether of nanoscale materials or poetic texts, is that activity within which working and thinking, intellectual and manual labor, theory and praxis, *technē* and *poiēsis* fuse into "skilled making."[15] Within the post-Fordist regime of accumulation of which *Mad Science in Imperial City* is a product, thinking and acting (in general) are increasingly functionalized as work. In other words, one way in which the ideological free-indirect discourse of Wang's formulation—"work nano, think cosmologic"—transforms the earlier imperative to "think globally, act locally" is by converting an ethical and political imperative into one that specifically concerns *labor* and therefore involves us with the critique of political-economy.

Two questions, then: How does the imperative to "work nano, think cosmologic" shift the problem of scale at issue for cognitive mapping? And how can we understand and historicize that imperative's reference to working and thinking in terms of the shifting status of the division of intellectual and manual labor?

To begin with the second question, we can turn to Alfred Sohn-Rethel's effort to understand the capitalist division of intellectual and manual labor in terms of *real abstraction*, the structure of the exchange relation at the root of what he terms the "social synthesis." In *Intellectual and Manual Labor*, Sohn-Rethel offers a historical materialist critique of Kantian epistemology, arguing that the abstract forms of our cognitive relation to the world—what Kant called the conditions of possible experience—are correlated to and indeed produced by the structure of abstraction proper to the capitalist exchange relation. According to Sohn-Rethel, the structure of cognition is *genetically* related to the abstract formalism involved in the subsumption of use value by exchange. "Time and space," for example (the Kantian forms of intuition), are

> rendered abstract under the impact of commodity exchange. . . . The exchange abstraction excludes everything that makes up history, human and even natural history. The entire empirical reality of facts, events and description by which one moment and locality of time and space is distinguished from another is wiped out. Time and space assume thereby that character of absolute historical timelessness and universality which must mark the exchange abstraction as a whole and each of its features.[16]

Sohn-Rethel thus argues for the material-historical genesis of cognitive structures, a genesis constitutively conditioned by capital. He pursues this argument toward an analysis of the Marxian category of "real abstraction," the generation of a social bond that is material-abstract, operative in the globally synthetic determination of value by socially necessary labor time—a synthesis that structures the process of valorization and is condensed into every physical act of exchange. As Sohn-Rethel explains, the "real abstraction" that is at stake in exchange and valorization has a curious and contradictory double physicality at its root:

> We noticed that the exchange pattern of abstract movement has a peculiar contradiction at its root. In exchange, abstraction must be made from the physical nature of the commodities and from any changes that could occur to it. No events causing material changes to the commodities are admissible while the exchange transaction is in progress. On the other hand, the act of property transfer involved in the transaction is a physical act itself, consisting of real movements of material substances through time

> and space. Hence the exchange process presents a physicality of its own, so to speak, endowed with the status of reality which is on a par with the material physicality of the commodities which it excludes. Thus the negation of the natural and material physicality constitutes the positive reality of the abstract social physicality of the exchange processes from which the network of society is woven.[17]

Sohn-Rethel contends that these "two contrasting physicalities" correspond to a distinction between "primary nature" and "secondary nature": the first "created by human labor," the second "ruled by relations of property." Primary nature thus pertains to manual labor (the physical production of commodities), while secondary nature pertains to intellectual labor (ownership of property and management of the production process according to the exigencies of capital accumulation). Crucially, however, this second nature is not only that of ownership and management; it is also that of *thought* and of the abstract systems that it generates: those of philosophy and science. Through its separation from use value, commodity exchange produces an "abstract nature" that "is devoid of all sense reality and admits only of quantitative differentiation," and "this real abstraction is the arsenal from which intellectual labor through the eras of commodity exchange draws its conceptual resources."[18]

So how are we to situate the real abstraction within the period during which we move from "think globally, act locally" (the 1960s) to "work nano, think cosmologic" (2005)? Crucially, the period between the 1960s and the end of the twentieth century was marked by the transition to a post-Fordist regime of accumulation that renders the relation between manual and intellectual labor more complex. The managerial and technological innovations of Taylorism and Fordism not only enabled high profit rates during the postwar period; they also resulted in a tendential displacement of living labor power. Thus the historical accomplishment of what Marx calls "real subsumption" (the achievement of a properly capitalist production process) also necessitated a transformation in the composition of the working class and the nature of labor.[19] In *The Society of the Spectacle* (1967), Guy Debord summarizes some of the structural changes this portends:

> If automation, or for that matter any mechanisms, even less radical ones, that can increase productivity, are to be prevented from reducing socially necessary labor-time to an unacceptably low level, new forms of employment have to be created. A happy solution presents itself in the growth of the tertiary or service sector. . . . The coincidence is neat: on the one hand, the system is faced with the necessity of reintegrating

> newly redundant labor; on the other, the very factitiousness of the needs associated with the commodities on offer calls out a whole battery of reserve forces.[20]

We might add that technological and managerial developments enabling automation not only free up labor for the growth of the service sector but also produce the technical conditions of possibility for new modes of communication, intellectual labor, and valorization enabled by information technology.

The transition marked by Debord has been theorized by Maurizio Lazzarato as a shift toward "immaterial labor"—a term Lazzarato quickly (and wisely) abandoned because of the irresolvable problem of reconciling it with a materialist position.[21] The salient transition he meant to denote, however, was the increasing importance of the "informational content" and the "cultural content" of the commodity: the increasingly constitutive role of information technology to labor in both industrial and service sectors, and the subsumption of cultural activities not usually considered as "work" within the valorization process. For Lazzarato, "the great transformation that began at the start of the 1970s" involved the hybridization of intellectual and manual labor, conception and execution, labor and creativity into "mass intellectuality."[22] Whatever one thinks of Lazzarato's terminology, we should note that it is indeed the case that deindustrialization and the growth of the tertiary sector is a *global* tendency of the period between 1973 and 2000. The structural shifts theorized by Debord and Lazzarato indicate the dialectical character of the real abstraction, which both emerges from and transforms the process of valorization through the technical innovations that process breeds and the altered relations between working and thinking those innovations at once require and produce.[23]

In *The New Spirit of Capitalism*, Luc Boltanski and Eve Chiapello analyze the relation between management and work proper to post-Fordism, emphasizing a transition between the 1960s and 1990s from hierarchical models of organization into a "network" model wherein organization is "*flexible*, *innovative*, and highly *proficient*" and workers function in "self-organized" and "creative" teams. By the 1990s, Boltanski and Chiapello argue, we are dealing with

> an economic universe where the main source of value added is no longer the exploitation of geographically located resources (like mines, or especially fertile land), or the exploitation of a labor force at work, but the ability to take full advantage of the most diverse kinds of knowledge, to interpret and combine them, to make or circulate innovations, and, more generally, to "manipulate symbols."[24]

This is the "economic universe" to which a slogan like "work nano, think cosmologic" pertains. In such a universe, it is adaptability, flexibility, and versatility that are the most highly valued qualities of workers, the ability "to switch from one situation to a very different one, and adjust to it" or the capability of "changing activity or tools, depending on the nature of the relationship entered into with others or with objects."[25] Intellectual and manual labor thus blend into the "virtuosity" of deploying flexible "skill sets" and technical capacities: precisely the kind of virtuosity that Wang's transition from engineer to poet exemplifies.[26] Boltanski and Chiapello figure the shift from hierarchical to network models, and also from the rigid division of labor to "flexible skills sets," in terms of a shift from the industrial city to what they call the "projective city." In the projective city, they argue, it is *activity* that serves as the general equivalent, the abstract standard by which persons and things are measured:

> In contrast to what we observe in the industrial city, where activity merges with work and the active are quintessentially those who have stable, productive waged work, activity in the projective city surmounts the oppositions between work and non-work, the stable and the unstable, wage-earning class and non-wage earning class, paid work and voluntary work, that which may be assessed in terms of productivity and that which, not being measurable, eludes calculable assessment.[27]

This is the "social synthesis" of what they call "the new spirit of capitalism" during which sharp distinctions between manual and intellectual labor, working and thinking are increasingly elided and the boundaries of what constitutes "work" become more and more permeable.

If the division of intellectual and manual labor is thus to some degree undermined, how has the relation between the real abstraction of the exchange relation and the nature of conceptual abstraction been transformed? In *What Should We Do with Our Brain*, Catherine Malabou analyzes parallels between "the new spirit of capitalism" described by Boltanski and Chiapello and transformations in our understanding of cognition over the past twenty years. She shows that cognitive models based on centralized computation have given way to network models of neuronal organization emphasizing the decentralization of cognitive functions and the distributed, adaptable, multifunctional action of neuronal "assemblies."[28] Contemporary neurology, Malabou argues, emphasizes the *plasticity* of the brain, understanding cognitive structures as both *formative* and *formable* (in contradistinction from the immutable transcendental structures of cognition analyzed by Kant).[29] As Malabou points out, Boltanski and Chiapello themselves note the parallelism of this cognitive model with the character of post-Fordist accumulation, arguing that "neuronal functioning and social functioning interdetermine each other and mutually give each other form . . . to the point where it is no longer possible to distinguish them."[30] In other words, the relation *between* neurological research and the regime of accumulation within which it is embedded is itself dialectically "plastic" in the sense described by Malabou. Thus the structure of the real abstraction moves with the moving contradiction between capital and labor, and so do theories of the structure of cognition. Such would be a historical materialist account of abstract thought following from Sohn-Rethel's critique of idealist epistemology. Put otherwise, not only *working* but *thinking* undergoes structural alterations conditioned by the history of capitalism, by historical mutations of the contradiction between capital and labor.

If, as I have argued, the phrase "work nano, think cosmologic" both mimics the glib language of promotional technobabble *and* expresses a genuine effort to include vastly discrepant orders of scale within the range of one's experience, that is because it is not at all simple to distinguish the ideological imperatives that interpellate us as subjects from our real subjective desires. Wang cannot escape the *mediation* of how he thinks by the context in which he works, a mediation that renders these two things inextricable. The real abstraction is the *medium* of thought, in which forms of thought both emerge from labor and occlude their relation to its physicality. The post-Fordist context in which working and thinking are increasingly hybridized into vaguely demarcated forms of "activity" is indexed by the saturation of Wang's poetic writing by scientific knowledge: he quits his job as an engineer so as to write poetry suffused with the cognitive tools of his former trade.

The import of this relation is signaled on the first page of Wang's book, a description of a nightmare in which the speaker attempts to read a poem titled *Terracota Warriors Quartet*, then switches to a conference paper titled *Nanoscale Prototyping of Anthropoid Integrated Circuits*. Immediately, we have an artist and a scientist whose work is his thought and whose medium is the virtuosity of both performance and adaptable skill sets. But as the speaker mumbles and misplaces his manuscript, the conference room becomes a music hall, and the halting narrative of the nightmare fractures into dreamlike memory work:

> Across River Y that pours down from heavenly sky, every early July we boarded the train home from the ancient city; midway I changed to another while he continued on to the capital. But the music never discontinued. I am the standing waves confined in

> the delicate metal strings of his violin, the tensions and therefore frequencies of the notes which are measured by four characteristic levels, *Aut, Win, Spr, Sum*, respectively, according to the seasons in the ancient city. It follows from well-established fact that elastic material undergoes thermal expansion or contraction when temperature fluctuates. (*MS* 3)

Autumn, Winter, Spring, Summer: perhaps we are reading a seasonal poem, a georgic in four parts, tracking our works and days across the movements of the year. But this is a georgic in which the river's name is denoted algebraically, the speaker identifies himself in the language of quantum mechanics, and seasonal changes of temperature are described through the physical properties of elastic materials. Indeed the seasonal structure of *Mad Science* is implied by its division into four sections—"PVD," "J Integral," "A's Degeneracy," and "T Square"—but these are titles that immediately torque the classical and romantic precedents of the seasonal poem into the world of materials science, mechanics, quantum physics, and engineering. Given this maneuver, we should recall that the seasonal poem itself has always been closely bound to the relation between intellectual and manual labor: to agriculture, to the idealized life of the "pastoral" worker and the beasts of the field, but also to the representation of these by an articulate poetic voice bespeaking the rhythmic harmony of the natural and the artificial. My argument is that in moving from "think globally, act locally" to "work nano, think cosmologic," Wang both resituates the 1960s idyll of the global village at the outer limits of scale and insists upon a consideration of the relation between intellectual and manual labor as crucial to poetic writing. He thus situates us both at the outer limits of "nature" and within the movement of recent history, taking up the contemporary inextricability of work and thought as the most intimate materials of his writing.

MAD SCIENCE

We can get a better sense of how Wang incorporates the scientific materials of his past occupation into his intellectual labor as a poet by tracking the four subtitles of his book across the referential fields they draw together. Doing so will begin to elucidate his poetics in more detail.

The first section of *Mad Science* is titled "PVD": in materials science, an acronym for "physical vapor deposition," a means of depositing thin films on a surface through the condensation of a vaporized material. In Wang's book, PVD refers to "Physical Voice Deposition" (*MS* 10), a process that seems to be crucially bound up with the problem of writing poetry in a second language:

> The gaseous suspension of particles, the voice particles. Are they produced by vaporization of or through impact by high-energy foreign ions on parts of speech in my utterance? What and where are the energy sources? (*MS* 10)

The suspension of voice particles is produced by either the vaporization of one language or its collision with another: "impact by high-energy foreign ions on parts of speech." As is often the case in Wang's writing, the defamiliarizing difficulty of syntax and grammar is manifest in the knowingly awkward construction of the sentences ("parts of speech in my utterance"), and this difficulty puts the process of "physical voice deposition" into relation with the problem of origin that always haunts the relation between speech and writing. "What and where are the energy sources?" the speaker asks. As the 'pataphysical framing of these questions implies, the question of "energy sources" is one not only of language or nationality but also of the relation between science and poetry, *technē* and *poiēsis*, the double sources of Wang's work. Physical Voice Deposition is not only the thin film that draws together speech and writing, or one language and another, but also knowledge and inscription, work and thought, science and poetry. It is the process by which the relation between these is *fabricated.*

"J Integral," the title of the second section, refers to an equation that allows one to calculate the strain energy release rate (or work) for a crack in the surface of a material. In *Mad Science*, "J Integral" is a two-page prose poem in which an equation figures the work of memory, and the work of mourning, through which the speaker's fractured recollections are configured and composed. "J is short for *Jiaotong* in my mother tongue" and therefore connotes Xi'an Jiaotong University, where Wang studied engineering. At Jiaotong, the speaker in *Mad Science* "walked every week with K and X" along a path circumscribing a "quiet man-made lake in the park of the ancient city," a path that becomes "the line integral along the curve *J* in the *K-X-I* space, $\int_J PdK + QdX + RdI$ which perfectly approximates the accumulation in my memory of our hundred-time walk" (*MS* 20). Narrative characters are alphabetic characters, K, X, and I, functioning as variables in the equation. Here, the "*K-X-I* space" is a region of memory, and the speaker tries to solve the line integral (the sum of values of a field at all points of a curve) of the hundred-time walk this region includes. As the speaker makes clear, it is a region that also includes a fracture, surrounded by "the curve *J*": "At 4:30 a.m. on Monday X phoned me that K had died unexpectedly while playing his violin in Stockholm. Two days later X fled the city too and secluded herself in a remote monastery" (*MS* 19). The death of one friend and the disappearance of another is a fracture in the surface of a life, a crack within a region of memory figured as the "*K-X-I* space." The J Integral calculates the strain energy release rate at the site of this fracture, a quantitative measure of the work of mourning.

The third section of *Mad Science* is titled "A's Degeneracy," with twenty-five subsection titles reconfiguring the phrase "HOW TO WRITE" (e.g., "HOW WRITE TO," "TO WRITE HOW," "HOW TO WRITE TO," etc.). In physics "degenerate matter" refers to matter so highly compressed that its pressure is due not to temperature but to the intense proximity of particles that can neither change energy levels nor occupy identical quantum states. Degenerate gasses, such as those in collapsed stars, are extremely dense and resistant to further compression. Their particles are so close together that they cannot be closer, nor can they change location. In *Mad Science*, the section titled "A's Degeneracy" opens in an implicit quarrel with a poetry instructor's commitment to Poundian modernism, a poetics foregrounding the maximal compression of concrete images:

> His obscure remarks on that little yellow note stuck on my 1st poem, You can use the word *W* . . . in your . . . Poetry is not for you, A. I was told that he was referring to relying on concrete images and metaphors and resisting abstraction in poetry. But I had many grudges against this blasphemous affair between the Deity of Poetry and earthly images and metaphors. (*MS* 25)

"*You can use the word . . . W-O-U-N-D . . . in your-poems-only once-in your lifetime*," we read later; "*everything-must be-concrete . . . photographic images, no abstract words, no freedom, no truth, no justice, no soul, no heart, no love*" (*MS* 99). The Poundian poetics of *condensare* relates the desideratum of concision, compression, to the privilege of the concrete over the abstract. "Use no superfluous word," Pound counsels, and "go in fear of abstractions." "Don't use such an expression as 'dim lands *of peace*.' It dulls the image. It mixes an abstraction with the concrete."[31] The abstraction is extraneous to the expression, and its excision makes the line *compressed* by making it *concrete*. Wound is a word, however, that Wang cannot only use once, because "my wounds have never been singular as shown by the X-ray images, and I can't spell any one without mentioning all of them at once" (*MS* 99). The abstract noun is so replete with the concrete significance it compresses—so dense with the pressure of collapsed contingencies—that to use it once is to unleash a multiplicity demanding reiteration. For Wang, there is no natural object that can serve as a concrete image of his wounds. "Words arrive in triangles and squares and leave in circles without any lasting registration in the brain," he writes. "However my wounds just stay. They never came. I never let them go. My abstract wounds with no definite shapes and locations" (*MS* 97). The speaker's wounds are iterable and necessarily reiterated because they have no determinate location, no single context within which they can be delimited, and they are abstract for the same reason: they evade the condensation of the concrete image, they evade *figuration* because they are without figure, dis-figured, without origin or end, refractory to the present and to presentation. If the density of degenerate matter is such that it resists further compression, the traumatic matter of Wang's poem is already so highly condensed, the wounds that it bears so proximate, that the written treatment of these requires expansion rather than contraction, distributed repetition rather than condensed concision, the generality of the abstract rather than the singularity of the concrete.

The final section of *Mad Science* is titled "T Square," an instrument used to draw and measure straight lines on a drafting table. As the title of the final section of Wang's book, "T Square" draws together this obedience to the *rule* of drafting or measure with references to Tiananmen Square, to Times Square, and also Times' Square, as the speaker puts it earlier in the text: "the crucible of infinite duration of forgetting" (*MS* 82). The Square, in Beijing and in *Imperial City*, is "the capital within the capital":

> The purple streets for a grid with a 4-fold broken rotational symmetry about the axis of the bell tower. It's a blood-baptized conformal (angle-preserving) mapping of the ancient city. All my study about T Square revealed no hint of this underground city. But there has never been any city but this one. (*MS* 121)

In *Mad Science*, the imperial city becomes any-city-whatever, New York or Beijing, as the repression at Tiananmen comes to map the space of a global world order within which there is nowhere else to go, however far one travels. The T Square is thus always here, anywhere the speaker is, yet it is also irreducibly *there*, a place where one has been, that one has left, and to which one cannot return. It is the pathos of longing for a place, which is also a palpable dread at the placelessness of power, that lends Wang's text its excruciating affective torque:

> I square and square more questions, imaginary questions like do you still possess your old T-square, is the T Square a blown-up self-adjoint projection of the T-square onto the capital, or the T-square a miniaturized prototype of the T Square.
>
> As if for the Q & A session following my presentation.

> I square root, again and again, these complex questions, which boil down to the one and only question, with real part, *what will you miss*, imaginary part, *if you are never able to walk in the T Square again*.
>
> The bell rings and rings.
>
> I drift in the complex conjugate of the ultimate question, that is, *what will you miss if you are able to walk in the T Square again*, circling the inverse square (laws of electricity and gravity).
>
> I look back at the invisible Square.
>
> Will you ever walk again in the T Square or not?
>
> I watched the inverse of the sparse square matrix burn at dawn. (*MS* 127)

"It's such a capital sin of the body not to be able to drown in the ruins" (*MS* 127), writes Wang. But the body does not drown; like the ruins, it remains. All that the body can do against its failure to dissipate is inscribe these questions in writing.

What was Tiananmen? What was 9/11? What is "the inverse of the sparse square matrix"? How do we get from Tiananmen Square to Times Square, from Beijing in 1989 to New York in 2001? How is one the inverse of the other? According to Wang Hui's account in "The 1989 Social Movement and the Historical Roots of China's Neoliberalism," "the 1989 social mobilization was criticizing the traditional system, but what it encountered was not the old state, but rather a state that was promoting reform, or a state that was moving gradually toward a market society."[32] Tiananmen was the crux at which totalitarian state socialism intersected China's integration within the market mechanisms of global capitalism. The repression of June 4, 1989, was the hinge upon which this transition turned, securing the terms upon which it would take place as those of state power. September 11, or what the state would make of it (the Patriot Act, the invasion of Iraq), was the totalitarian response of the United States to the waning of its neoliberal hegemony, the end of the conditions of possibility upon which the pretense of a *pax Americana* upheld America's position within the capitalist world order. The recursion of 2001 upon 1989 effects an inversion through which the end of history ends, and the hybridization of neoliberal capitalism and totalitarian state power expresses itself one way or the other. "The world is upside-down," writes Wang; "the world is the mirror image of my wounds" (*MS* 89).

Within this inversion, Wang attempts to transcribe the *somewhere* of a wound within the *here* of a body attached to the *there* for which it yearns and situated within the *anywhere* of a power from which it cannot escape. The "T Square" described by the speaker is "coated with carbon nanotubes" (*MS* 113); it is "almost frictionless and thus wear resistant." The place of politics—the public square—resists resistance and public assembly, such that "sliding is the only form of motion, and the only way to change direction is through impact against the wall" (*MS* 113–14). Describing a "nanostructured greatcoat which resists diffusion of any kind through it. All invisible to normal human vision," the speaker remarks that "we have finally materialized the cloth of the emperor" (*MS* 115). Here, advances in materials science invert the function of ideology in the story of the Emperor's New Clothes. It is not the case that the Emperor's real powerlessness, the self-evidence of his nudity, is covered by an imaginary supplement or occluded by ideological disavowal. Rather, the Emperor's apparent nudity is itself illusory, concealing the insensible presence of a material power whose effectivity can neither be denuded by critical consciousness nor breached by oppositional force. Nanostructured materials serve as a synecdoche for the operations of an insensible power that cannot be "seen through" because it is not seen. What the critique of ideology cannot denude is the reality of the real abstraction,[33] the material-insensible social bond through which the general intellect is applied, for example, to the production of military technologies. The routing of technological innovation toward the maintenance of class power and imperial domination is an index of the real abstraction's insensible effectivity, mediating the relation between work and thought in such a way as to enable the reproduction of the social synthesis. As usual, Wang situates the social effectivity of the real abstraction in the most expansive context: "Dark energy. White momentum. Gray matters of the brain. Gray interior of the A train. The Andromeda spirals away at ever-increasing angular velocity. What an abstracted afternoon" (*MS* 97). From Tiananmen to Times Square to the gray interior of the A train, among workers on their way home, it is the gray matter of the brain—the speaker's body, the material substrate of his thought—that carries the wounds it must transmit across geography and history, somewhere within the angular spiral of a galaxy.

THE ENCOUNTER, THE COLLECTIVE

Dark energy and gray matter but also "Dark matter. White page" (*MS* 36): insofar as they enter into poetry, abstractions are *inscribed*, and inscription is thus the site of a material mediation between the abstract and the concrete. Under the subsection title "HOW TO WRITE" we find an image of a crumpled page from the *American Heritage Dictionary* (4th ed.) upon which we can barely make out the definition of the word "abstract" (Fig. 58). Insofar as it is written, the abstract must be concrete: it must have a material substrate, for example. And insofar as it is written, the concrete must be abstract: it must be iterable, reproducible, contextually mobile.[34] Immediately below this scanned definition of the word "abstract" we find the passage that also serves as an epigraph to the prologue of the present book:

> In the class-A clean room of the National Nanofabrication Laboratory in the heart of Silicon Valley, under the ultrahigh-magnification AFM (Atomic Force Microscope), I thumbed through every page of the whole collection of poetry books stolen from the Public Library, and I found only dots, dotted straight lines, dotted arcs. Abstract geometrical entities banging my dilated eyeballs. (*MS* 26)

Here, the "ultrahigh-magnification" of nanoscale microscopy converts concrete signifiers into abstract geometrical forms and these, in turn, assail the eyeballs as "geometrical entities," reminding us of the physical instantiation of "the senses" while rendering the sense of the poetry illegible. It is precisely this *illegibility* with which Wang is concerned: the unreadable concreteness of the abstract, the untranslatable abstractions we encounter at the core of the concrete, the chiasmus of the "abstract" and the "concrete" through the crux of that which has to be written yet cannot be read: "my abstract wounds with no definite shapes and locations." In *Mad Science in Imperial City*, the real abstraction is the paradoxical site of an unplaceable wound: it is at once the form and genesis of the *way we think* and thus the way we might come to terms with history. But since it is the way we think, it is that which cannot be thought, a ruse of reason. The wound Wang wants to think (history) is that which—to deploy Lacan's formulae for contingency, necessity, and impossibility—has "stopped not being written" (contingency), that "doesn't stop being written" (necessity), yet which "doesn't stop not being written" (impossibility).[35] The implacable impossibility of writing the wound is precisely that it binds together contingency and necessity, rendering a contingent event inescapably *written* and yet inescapably *unwritable*. "It simply happened," Wang writes, "an event without answer" (*MS* 121). The wound, what simply happened, is the inscription of a body in history, an event that remains unanswered insofar as history *goes on* inscribing itself, carrying the body with it.

"Something in the world forces us to think," writes Gilles Deleuze (*Il y a dans le monde quelque chose qui force à penser*).[36] And "this something," he continues, "is an object not of recognition but of a fundamental *encounter*."

HOW TO WRITE

Yes, I say. This is what I found in my *American Heritage Dictionary* (4th Ed.) through the long summer evening.

In the class-A clean room of the National Nanofabrication Laboratory in the heart of the Silicon Valley, under the ultrahigh-magnification AFM (Atomic Force Microscope), I thumbed through every page of the whole collection of poetry books stolen from the Public Library, and I found only dots, dotted straight lines, dotted arcs. Abstract geometrical entities banging my dilated eyeballs.

Yes, I say. Last night I had this dream. I the 3rd walk with my mother and the flock of my siblings, half-siblings, unknown siblings to a concert in the ball park named after an internet tycoon on the edge of the industrial park to hear old songs before her returning home across the ocean after a brief visit. I weep before the opaque wall of cacophony because I've suddenly noticed that she still looks incredibly young and beautiful and short after all these years she has lived through alone, because she has

26

Figure 58 The concrete definition of "abstract." Shanxing Wang, *Mad Science in Imperial City*. Courtesy of Shanxing Wang.

What is encountered is that which "can only be sensed" (*Il ne peut être que senti*), insofar as it cannot be remembered, thought, or recognized. Accessible only by way of the contingent encounter, it has to be *felt*. But insofar as it can only be sensed or felt, rather than recognized, the object of the encounter is also *insensible*: insensible because what is encountered is "not a sensible being but the being *of* the sensible."[37] The element of the encounter is thus *intensity*, understood as pure difference in itself: not the given, but differential intensity as the element through which the given is given.[38] That which, in the world, forces us to think is the insensible differential that can only be contingently encountered in and through the sensible.

The problem of writing the wound, "the event without answer," is the problem of writing the encounter. The object of the encounter evades figuration, insofar as it is an object that evades identity, recognition, resemblance, or analogy. The speaker's wounds are "lithographic witness of the pretentiousness of similes. Atomic testimony to the impossibility of metaphors" (*MS* 60). Thus he finds himself confronted with the materials of writing, with the contact of inscription:

> this sentence is a visco-elasto-plastic line this line walks runs pauses leaps turns and breaks grudgingly into multiple lines of even length forced by the hard walls of the page this line is the track left by the sliding and rolling contact of the metal ballpoint on this 8.5 X 11 black plane of recycled carbonless paper this moving contact is transient thermal history-dependent and irreversible. (*MS* 130)

Because the wound evades *figuration*, Wang probes the *configuration* of the materials but finds that what he is looking for is not quite here either. "There are no gaps in the *real* line," the speaker writes, but "this line is not a *real* line because in it gaps abound gaps of irrational shapes of irrational numbers of irrational longings I write *this* as he says *that*." Irrational shapes, irrational numbers, irrational longings interrupt the material recording of the encounter, whether through the sense of the line or its sensible inscription. "This longing," Wang writes, "can not be spelled in any natural language this longing has no metric scheme this longing renounces punctuation" (*MS* 131).

But Wang's approach to the insufficiency of natural language and poetic figuration is not simply to gesture toward the unsayable. Nor is the difficulty exactly that of presenting the unpresentable. Rather, the problem is how to present the problem itself—how to *formalize* the irrationality of a gap that cannot be said and how to approach the longing to say it, in writing. Here, as does Deleuze in attempting to think the synthesis of difference,[39] Wang turns to mathematics, and specifically to calculus. If the speaker's longing "can not be spelled in any natural language" and if it "renounces punctuation," perhaps *the problem of how to inscribe it*—the problem of "HOW TO WRITE"—can be formalized in the common language of the nano and the cosmologic, mathematics:

> But this feeling is real this feeling is a real-valued function of the unreal time *t* denoted by *F* which is disintegrated but still perversely continuous and continuous to the order of ∞ this function spans the entire family of infinitely differentiable real functions in Euclidean space thus supplying the test function for the rigorous definition of the Dirac distribution $(\delta, F) = \int_{-\infty}^{\infty} \delta(t)F(t)dt = F(0)$ which is a continuous linear functional that smoothes out the δ singularity of the function and which always assumes the real value of $F(t)$ at origin that is the instant of this once-in-life intimacy with catastrophe. (*MS* 131)

The crucial reference in this passage is to the theory of distributions (discovered by French mathematician and political activist Laurent Schwartz) and to the application of distributions to the "δ function" (or delta function), formalized by the physicist P.A.M. Dirac, a major figure in the development of quantum mechanics. Briefly, Schwartz's discovery of distributions in 1944 solved a crucial problem in analysis by enabling mathematicians to work with functions that cannot otherwise be differentiated at certain exceptional points: the Dirac δ function, for example, represented by an infinitely sharp peak (or "singularity") with a value of 0 everywhere except at its origin, where its value is infinite. The theory of distributions invented by Schwartz allows one to treat such a function through differentiation. As Wang puts it, distributions enable one to "smooth out the singularity of the Delta function" for analysis, as in the graphic representation on the cover of *Mad Science in Imperial City* (Fig. 59).

In engineering, the Dirac distribution enables one to approximate the force of an instantaneous load or an "impulse" that theoretically involves no temporal duration ("the unreal time *t*," in Wang's text). Thus, while it is an equation that Wang would have worked with regularly as a mechanical engineer, here it enables an approach to "the instant of this once-in-life intimacy with catastrophe." It inscribes an approach to that *feeling*, denoted by *F*, of a longing that can not be spelled in any natural language, that has no metric scheme, that renounces punctuation, that renders similes pretentious and metaphor impossible.

But in what sense is this not a metaphor ("this feeling is a real-valued function")? How can the 'pataphysical model of the Dirac distribution, by which the speaker's feeling is approached, avoid reinscribing his unsayable longing within

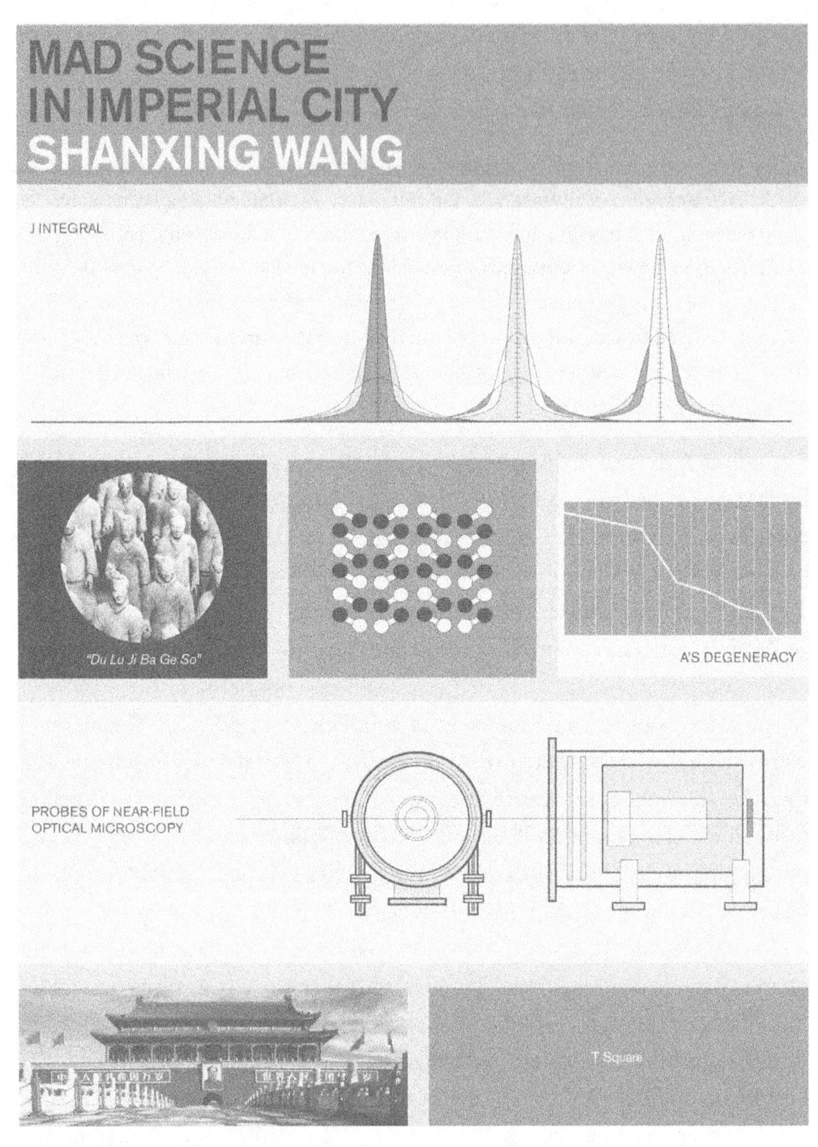

Figure 59 Cover of Shanxing Wang's *Mad Science in Imperial City.* The top panel shows the operation of "smoothing out" the singularity of the Dirac distribution. Courtesy of Anthony Monahan.

the regime of analogy, which cannot account for the encounter? Indeed, how is it that the entirety of *Mad Science in Imperial City* does not circumscribe the writing of memoir within a system of analogies between historical experience and scientific discourse?

In "Sincerity and Objectification" (1931), Louis Zukofsky cites William Carlos Williams's critique of "crude symbolism" in *Spring and All* (1923):

> Crude symbolism is to associate emotions with natural phenomena, such as anger with lightning, flowers with love; it goes further and associates certain textures with . . . [*sic*]. It is typified by the use of the word "like" or that "evocation" of the "image" which served us for a time. Its abuse is apparent. The insignificant "image" may be "evoked" never so ably and still mean nothing.[40]

As evidence that the simile "can be not a wandering ornament, but a confirmation of the objects or acts which the writer is setting down," Zukofsky cites two lines by Charles Reznikoff:

> From the bare twigs
> Rows of drops like shining buds are hanging.

"I.e.," Zukofsky remarks, "each drop is close to where a bud may be."[41] It is the plausible spatial *proximity* of drops to buds that motivates the figure and thus makes plausible the simile itself. Zukofsky finds that Reznikoff's simile is justified because it attests to an "accuracy of detail in writing" that he terms *sincerity*.[42] It is in this sense that we may also understand the significance of the term "sincerity" in Wang's epigraph to the final section of *Mad Science*, from Peter Handke: "The merit of originality is not novelty but sincerity" (*MS* 112). "This feeling is a real valued function" is not a figure motivated by analogical thinking because it is the inextricability of Wang's education as an engineer from the political catastrophe of Tiananmen and from the task of poetic writing that renders the equation an apt inscription of a problem: precisely the problem of how that inextricability is manifest as a feeling. Because it binds together working and thinking with politics and history, the equation functions poetically as a *formalization* binding the complex relation of these elements in the written signifier. The equation records not the substitution for, resemblance between, or identity of one term and another; rather it formally registers the feeling of their complex in a material inscription. The tools of a trade that Wang learned at one time are now available to him at another time, for use in another context: it is the feeling of the interval that is written.

It is not only a feeling that is formalized in Wang's inscription of the rigorous definition of the Dirac distribution; it is a poetics. The kind of "sincerity" inherent to the originality of this passage is in some respects similar to that which we find in the first poem of Williams's *Spring and All* ("By the road to the contagious hospital"). In the latter poem, as in Wang's use of the Dirac distribution, it is material gleaned from an extra-poetic profession that lends the piece its "accuracy of detail in writing." Whereas it is Wang's work as an engineer that renders his use of mathematics sincere within the context of his writing, it is his familiarity with the de-romanticized mise-en-scène of the maternity ward that lends Williams's carefully indirect construction of the relation between "birth" and "spring" its power of defamiliarization. But consider, also, the difference between these examples. The poem by Williams trades upon the occluded image of a mottled infant being born and the implied relation of this image to the approach of "sluggish / dazed spring" in a landscape of "standing water" and "reddish / purplish, forked, upstanding, twiggy / stuff of bushes and small trees."[43] Although the complex of the image is recessed and dispersed throughout the poem, the poem demands of its reader that the image be grasped and reconstructed. What we encounter in Wang's poem, however, is the formalization of that for which there is no image—not even an occluded or fragmented image. Where the image fails, there mathematical inscription intervenes. What has to be understood, in order to grasp the poetic significance of Wang's use of the Dirac distribution, is the significance of this intervention.

Recall my critique, in the introduction, of Daniel Tiffany's account of the relationship between scientific materialism and lyric poetry. For Tiffany, poetry's participation in the fabrication of the real depends upon its power to formulate *images* of invisible material substance, of primary corpuscles refractory to empirical observation. The lyric image fills out the lacuna of objectivity, insofar as scientists, philosophers, and poets evoke otherwise insensible matter through figuration. "When materiality is equated with invisibility," writes Tiffany, "the invisible world becomes the province not only of atoms and animals but also, conceivably, of beings possessing the radiant body of an angel."[44] What Tiffany's account overlooks, however, is the crucial role played by mathematics in modern science. For the scientist, it is mathematical formalism, not lyric figuration, that enables cognitive access and lends descriptive consistency to configurations of matter below the threshold of the visible.

In order to accord the image its role, Tiffany relies upon a preliminary alignment of scientific materialism with empiricist epistemology. This alignment generates what Tiffany describes as "an irresolvable conflict between ontology and epistemology within atomist doctrine: only atoms truly exist, yet, according to the premises of empiricism, they are unknowable and indeed inconceivable."[45] What this alignment of materialist ontology and empiricist epistemology overlooks is precisely the role of *rationalism* and *mathematical formalism* in the construction of scientific knowledge. In modern physics—and particularly post-classical physics—it is not so much the poetic image that compensates for lacunae of observation but rather the rigor and relational consistency of mathematical formalisms. Consider Gaston Bachelard's claim, which he is at pains to establish throughout his epistemological writings, that "to think scientifically is to place oneself in the epistemological terrain which mediates between theory and practice, between mathematics and experiment."[46] For Bachelard, this epistemological terrain, mediating between mathematics and experiment, is characterized by a dialectical relation between rationalism and empiricism. "Empiricism and rationalism in scientific thought," he writes, "are bound together by a strange bond, as strong as the bond which joins pleasure and pain. Indeed, the *one triumphs by assenting to the other*: empiricism needs to be understood; rationalism needs to be applied."[47] It is the consistency of rationalist formalism that makes possible the rigorous construction of experimental conditions; experimental results are understood and made accountable to theoretical consistency through mathematical formalization. Moreover, it is precisely *the image* that this dialectic of rationalism (mathematics) and empiricism (observation) enables science to overcome. "Intuitions are very useful," writes Bachelard, "they serve to be destroyed": "by destroying its original images, scientific thought discovers its organic laws. . . . The diagram of the atom proposed by Bohr a quarter of a century ago has, in this sense, acted as a good image: there is nothing left of it."[48] What Bachelard calls "the philosophy of no" is the method through which science negates intuitive images of matter, which serve precisely as productive material for these negations. It is through mathematical formalism, its empirical application, and its rational reconstruction that science overcomes those images or figures that stand in for what we do not yet know.

I do not mean to argue for any direct equivalence between Bachelard's epistemology and Wang's poetics. The use of a mathematical equation to formalize a "feeling" is by no means the same as its use in physics or engineering, nor is the movement between experience and formalization in the production of scientific knowledge equivalent to that which may be involved in the writing of poetic memoir. What, then, do we gain by recognizing that the role played by mathematical formalism in Bachelard's "philosophy of no" may be more helpful in thinking through the relationship between scientific epistemology and poetics in *Mad Science* than the role played by the image in Tiffany's account of lyric substance? To begin with, grappling with the mathematical displacement of the image in Bachelard's epistemology forces us to grapple with the *problem* that Wang confronts in his writing: the relation of mathematical formalism to the failure of figuration to fill in the lacuna of the sensible. The equivalent of Keatsian negative capability implicit in Bachelard's "philosophy of no" is what Bachelard calls "applied doubt": not a universal Cartesian doubt but rather a doubt which, confronted with each particular object, requires the constitution of a particular *problematic*. In Bachelard's sense, to construct a problematic is to construct a program of research leading from the rational (already constituted formalizations) through the empirical (experimentation) to new formalizations (the rational construction of the object). This is what Bachelard calls "applied rationalism."[49] The "wound" that Wang seeks to bind through inscription, or the feeling that follows from its aftermath, is an object of such "applied doubt." It pushes Wang to traverse an argument with Imagist *theory* ("relying on concrete images and metaphors and resisting abstraction in poetry") through the experimental practice of his own writing, working toward a reconciliation of the abstract and the concrete through a poetics of the real abstraction: the intersection of working and thinking with history. The equations and diagrams that Wang deploys mark the point at which "abstract" thought is inseparable from both "concrete" labor and from the materiality of writing, from *inscriptions* that do not give way onto the phantasmagoria of the image. The "problematic" to which his wounds give rise is the insufficiency of "the natural image" of Imagist doctrine to the encounter that has to be written. "There must be another law," writes Wang, "for the field of the written" (*MS* 130). So the problem of *how to write* becomes the problem of finding formalizations that, without recourse to the image, are able to *displace* the immediacy of the wound, through exteriorization, while also *binding* it in a system of signification irreducible to natural language. This is the problematic within which Wang works, and the inscription of the "rigorous definition of the Dirac distribution" is not the solution but rather the point in the text where that problematic is precisely constructed and conveyed to the reader. When I asked Wang about his progress on a second book, he told me it would deal with his childhood in provincial China, growing up during the Cultural Revolution; then he added, deadpan, "and I'm really trying to figure that out mathematically."[50] Wang thinks mathematically, and this modality of *thinking* becomes inseparable from the *work* of poetic writing and from the feeling of the historical experiences that are written. It is this inseparability that is formalized by the equations inscribed in *Mad Science in Imperial City*.

What, then, does the specific phrase "work nano, think cosmologic" have to do with this problematic? In his discussion of the encounter, Deleuze distinguishes between "the contingently insensible" (that which is too small or too distant to be registered by the senses) and "the essentially insensible": not a sensible being that goes unperceived but rather an encounter with the being of the sensible. The nano and the cosmologic denote the boundaries of the contingently insensible, or that which is too small or too distant, while the relation between working and thinking inscribes the insensible conditions under which those boundaries are encountered. Politics and history are the medium not only in which Wang works and thinks but also the medium in which the essentially insensible encounter transpires. When Wang writes, "I escaped the massacre by a nanometer" (*MS* 35), he deploys the nanometer as a metric of the *contingently insensible*—that which is "too small" to be sensed. When he writes his feeling as "a real-valued function of the unreal time *t* denoted by *F*" he deploys the language of mathematics to formalize an affect that is *essentially insensible*. The relation between these two modes of writing constitutes "the epistemological terrain" of Wang's poetics. It is not only a materialist poetics but also a *historical* materialist poetics because rather than privileging the concrete over the abstract it concerns itself with the real abstraction: with the historically situated, insensible forms of thought through which one tries to write the mediation between collective history and the singularity of a life.

"I strive to piece together the wrecked *they* for *you*," Wang writes, and his poetry reminds us of how implicated "we" always are in the non-figurative power of number and measure:

> You are still walking. 20 millions of you. 10 millions of you. A million of you. 250,000 of you. 100,000 of you.
>
> You are still walking. {*1, 2, 3, 3,000, 300, 20,000, 3, 50,000, 1,000, 12, 100, 100,000 3,100, 200, 300,000, 1,000,000, 50,000, 1, 1,000, 4, 7,000, 241, 3, 2, 1, . . .*}
>
> Who are you? (*MS* 126)

"How many *I*s are necessary for the existence of an *us*?" the speaker asks elsewhere. "What are the latent heat of fusion and the melting point for the crystalline *I*s to become the liquid *we*? *When we meet again*, isn't it the way we used to sing?" (*MS* 42). The problem of the collective is a problem of number and a problem of scale. Here, this is also presented as a problem of fabrication: what is the latent heat of fusion? What is the melting point at which the crystalline *I*s will become a liquid *we*? *How* is the we composed? Wang's "Imperial City" is the hybrid place where history keeps posing the question of the constitution of the we, at once abstract and concrete, over and over again. In *The Century*, Alain Badiou situates this question as the primary problem we have inherited from the twentieth century:

> Today everything that is not already mired in corruption raises the question of where a "we" could originate that would not be prey to the ideal of the fusional, quasi-military "I" that dominated the century's adventure; a "we" that would freely convey its own immanent disparity without thereby dissolving itself. How are we to move from the fraternal "we" of the epic to the disparate "we" of togetherness, of the set, without ever giving up on the demand that there be a "we"? I too exist within this question.[51]

For Wang, also, this question of the relation between number, togetherness, and collectivity is the axis of iteration around which one century turns into another: "*When we meet again*," he writes, "isn't it the way we used to sing?"

It would be insufficiently dialectical to say that, through a poetics of the real abstraction, Wang is trying to write the primary qualities of history. Rather, he is trying to write the primary qualities of the secondary qualities of history: to formalize the feeling of his immersion in its movement. In this context, the effort to "work nano, think cosmologic" is at once a universalizing and rigorously particular effort to confront both the limits of reason and the limits of fabrication—and thereby to confront the real, in the sense conveyed by George Oppen:

> These things at the limits of reason, nothing at the limits
> of dream, the dream merely ends, by this we know it is the real
>
> That we confront.[52]

CONCLUSION: TECHNĒ, POIĒSIS, FABRICATION

The primary claim of this book is that *fabrication* is the common rubric under which we can situate the juncture of *technē* and *poiēsis*, and that grasping the relation between these two terms in this manner is constitutive of a materialist understanding of either one. *Technē* is not merely knowledge but also craft or skill; *poiēsis* can only be thought as letting-appear insofar as it is also concrete making. Not only at the intersection of these two categories but as their common determination, *fabrication* is skilled making. Fabrication means "to make anything that requires skill"; "to 'make up'; to frame or invent." As both a *material* and *discursive* practice, fabrication draws together working and thinking, intellectual and manual labor, in a constructive elaboration. To fabricate is to form or shape or manufacture, whether a legend, a lie, a poem, a metal stock, or a carbon nanotube.

To recall that a fabrication is a lie, an artifice, is also to grasp the practice of fabrication in a materialist way. A lie is only recognized as such when it fails, when it indicates its imperfection, when it implicates itself in the flawed surface of the real from the fault of which it is constructed. When it *reveals itself*, that is. Drawing together *technē* and *poiēsis*, fabrication draws together poetic *making* with technical *craft*, a doubling that entails duplicity. Heidegger thought these categories together on the ground of bringing-forth, revealing, rather than as making, manufacturing, or construction. To think these categories together under the rubric of fabrication and to understand the modality of revealing this involves as *artifice* is to ruin the authenticity Heidegger reserved for *alētheia* as the essence of *technē* and *poiēsis*, over and against the inessential practice of mere material making. It is also to ruin any sense of skilled making grounded in the adequation of material structures to ideal forms. It is to grasp this authenticity and this adequation, their inherent idealism, as a lie, and thus to grasp the practice of material making that belies them as fabrication. Recognized as artifice, *poiēsis* gives away its truth; rendered material, *technē* falls from perfect knowledge. Thought as fabrication, these adhere within "the aesthetics of the defect state," to deploy Geoffrey Ozin's phrase.[1] This is what Bernard Stiegler means when he refers to "the de-fault of origin" that displaces the priority of either man or technics, thought or construction, intellectual or manual labor, craft or making. "The technical inventing the human . . . the human inventing the technical": both the human and the tool are *made up*, fabricated in "the meeting of matter" by which each produces the other through a coevolution of cortex and flint, of writing, engineering, and thought.[2] Fabrication, in fact, is the single term Stiegler lacks to specify the "double plasticity" to which he refers, *the duplicity of invention* that binds the "who" and the "what."[3] Thinking with and against Heidegger, Stiegler shows that the essence of technology is indeed something technological, as is the essence of the human, Dasein. My claim is that the essence of both *technē* and *poiēsis* is fabrication and that "the human" is inessential, a fabrication. To be clear, I acknowledge that this is itself an inessential claim, insofar as it is *partisan*. It is not a truth claim, though it claims a certain kind of rigor. That is: it takes a rigorously *materialist* position on the determination of and the relation between *technē* and *poiēsis*. Fabrication is the duplicity of *technē* and *poiēsis* that we encounter at their limits.

However we approach the intercalated problems of form, craft, intention, determination, making, construction—in other words, *technē* and *poiēsis*—we always arrive back at a peculiar paradox that is constitutive of "invention": new forms, new structures, new devices, new materials are for the most part made by mistake. Yet it is precisely the directed, historically specific engagement with programs of fabrication, under the pressure of constraints, that generates the contexts in which productive errors can and do occur. This axis of avant-garde agency, experimental contingency, and material fallibility is where directed intention and material contingency collide. The Buckminsterfullerene was inadvertently discovered during experiments concerning interstellar plasma. Sumio Iijima produced cylindrical carbon nanotubes while trying to synthesize spherical Buckminsterfullerenes. Charles Olson writes about "the difficulties" in the first book of the *Maximus Poems*; but it is only the later dissolution of Olson's mythopoetic project into a broken collective of fragmentary occasions that makes those difficulties properly manifest. Ronald Johnson's unrealized intention to complete *ARK* with a "Dymaxion Dome over the whole" decompletes its architectonic framing. Christian Bök's "aesthetic of structural perfection" is intruded upon by the supplement of the occlusion, and even his effort to perfect the imperfect typography of the 1994 edition of *Crystallography* through an amended 2003 edition produces two discrepant versions of the "same" text, thereby violating the crystallographic logic of identical replication. Caroline Bergvall's performative misarticulations build slippage, distortion, and displacement into a process of decomposition. And Shanxing Wang describes the function of his poetry in terms of a politics of error. Perhaps the only way of accounting for what it really means for form to be "open" is through a thinking, which cannot help but fail, of this limit at which compositional agency is exposed to its own default: the recognition that every practice of skilled making is also a practice of making mistakes and that this duplicity is constitutive of fabrication. That would be another way of situating fabrication at its limits.

Olson developed a materialist poetics by recontextualizing Keats's notion of negative capability (the capacity to persist in doubts and uncertainties) in the wake of post-classical physics, considering the implications of the latter for a practice of poetic construction. I have argued that nanoscale materials engineering shifts the pertinence of indeterminacy from an epistemological or hermeneutic register to a pragmatic register, wherein fabrication reaches a scale at which the constitutive indetermination of matter has to be factored into concrete practices of construction. Perhaps the pertinence of poetry to what Daniel Tiffany felicitously calls "the fabrication of the real"[4] is not so much its capacity to produce images of what we cannot see but rather to abide within the indetermination of what we make. Whether intentionally or not, a line like Dickinson's "You there – I – here –" quietly *makes* legible the indeterminacy of sense that resides in the material construction of the line, not in the ambivalence of semantic reference. Perhaps the peculiar capacity of poetry, its capacity to open language upon its protosemantic dimension, is to demonstrate that our cognitive apprehension of indeterminacy via ambiguities of "meaning" participates in and is premised upon the indeterminacy of "the materials."

Like the singular plurality of Louis Zukofsky's term "the materials of poetry," the pluralization of the "limit" in the phrase "the limits of fabrication" already indicates the indeterminacy of the epochal closure to which that phrase alludes. K. Eric Drexler claims that nanotechnology will give us "nearly complete control over the structure of matter."[5] We should attend carefully to that "nearly," which suggests that the limits of fabrication may themselves be unlimited. In Whitehead's vocabulary, we can say that "'actual entities'—also termed 'actual occasions'—are the final real things of which the world is made up,"[6] as long as we acknowledge that "the community of actual things . . . is an incompletion in the process of production."[7] We *approach* the limits of fabrication, but as Louis Althusser reminds us in another context, the future lasts a long time, and "the lonely hour of the 'last instance' never comes."[8] As long as fabrication, at its limits, intersects with error, the limits at which we fabricate the real will remain *constructively* indeterminate. If the phrase "the limits of fabrication" points to the boundaries that our finite capacities encounter, it also suggests that "the Real / goes on forever."[9]

ACKNOWLEDGMENTS

The process of writing this book began when I was invited to join the design team for an exhibition on nanotechnology at the Los Angeles County Museum of Art. My thanks to N. Katherine Hayles for that opportunity and for her guidance ever since. Kate's exemplary mentorship at UCLA gave me the confidence to do this sort of interdisciplinary work, as did the inspiration of her pathbreaking scholarship. Michael North and Kenneth Reinhard had a formative influence on my understanding of modernist poetry and modern philosophy, and their discrepant approaches to intellectual inquiry helped me see things through different methodological lenses. Eleanor Kaufman and Sianne Ngai were discerning interlocutors and offered important feedback on several chapters. At Queen's University, Jed Rasula introduced me to the poetic tradition at issue in this book and encouraged me to pursue inventive approaches to theoretical reflection, while Philip Rogers instilled in me a sense of respect for the demands of literary interpretation. Earlier, Sue Barker, Gus Liadis, Don McCormick, and Greg Munce made it possible for me to begin thinking and writing in the first place.

I owe a great deal to Brian Rajski and Kate Marshall, who read and thought through every chapter of this book. Their intellectual companionship sustained me through years of research and writing in Los Angeles. For their friendship and intelligence, I am indebted to Vera Bühlmann, Kathleen Frederickson, Peter Hallward, Aaron Hodges, Amanda Holmes, Julian Knox, Alexi Kukuljevic, Rachel Kushner, Robert Lehman, Colin Milburn, Elizabeth Miller, Dee Morris, Marty Rayburn, Joshua Schuster, Jason Smith, Matthew Stratton, Stephanie Wakefield, Audrey Wasser, Evan Calder Williams, and Michael Ziser. Memorable afternoons of writing and discussion at the Getty Research Institute with Knox Peden made me a more stringent thinker, while Martin Hägglund's advice and encouragement helped me through the final stages of revision. Shanxing Wang generously discussed his work with me when I first read it, and his friendship has been an inspiration. Stephen Voyce has pushed me as a critic and supported me as a friend since I met him many years ago.

I learned a lot through seminars and discussions with a number of outstanding students at the University of California, Davis: Ian Afflerbach, Marlena Gates, Angela Hume, Thomas Johnson, Jasmine Kitses, Benjamin Kossak, David Rambo, and Bonnie Roy. At Concordia University in Montreal, I am lucky to have a context in which to pursue further research at the intersection of science, philosophy, and poetics through the Centre for Expanded Poetics and the resources of a Canada Research Chair. I am grateful to my dear friends Petar Milat and Tomislav Medak for inviting me to coordinate an ongoing series of annual philosophy symposia at MaMa Multimedia Institute in Zagreb, several of which have addressed conceptual problems at stake in this book. My thanks to Jacques Lezra and Paul North for initiating an inventive series like IDIOM and for including this book in it; I couldn't have asked for more intelligent editors or a better context for my work. Tom Lay, Richard Morrison, John Garza, Tim Roberts, and Jenn Backer expertly saw the book through the publication process. Thanks to Robin Graham for preparing the index.

Because she knows what matters and displaces what doesn't, Cynthia Mitchell made it possible for me to become the person who could finish this book while remaining the person who started it. Finally, I am grateful to my family for their love and support.

NOTES

PROLOGUE: LIMITS

1. The term "nano" denotes "a billionth," and the term "nanoscale" generally refers to scales between one nanometer (a billionth of a meter) and 100 nm. "Nanotechnology" refers to nanoscale instruments or to instruments capable of addressing or characterizing nanoscale materials. More broadly, it refers to technoscientific projects involving the controlled manipulation of nanoscale matter and the fabrication of nanoscale structures and devices. The more general term "nanoscience" refers to the *study* of matter and physical law at the nanoscale. Materials science, to which nanoscience and technology have made important contributions, is the applied study of the relationship between material structures and material properties, and the design and fabrication of materials with novel properties on this basis. For an introduction to nanoscience and technology for the general reader, see Wilson et al., *Nanotechnology*. On the cultural implications of nanotechnology, see Hayles, *Nanotechnology*. On materials research and fabrication, see Callister and Rethwisch, *Materials Science and Engineering*. For an introduction to developments in materials science made possible by nanotechnology, see Ball, *Made to Measure*. For a more technical presentation, see Cao, *Nanostructures and Nanomaterials*.

2. Feynman, "There's Plenty of Room at the Bottom," 360.

3. Ibid., 363.

4. On Newman's demonstration, see Regis, *Nano*, 143–47. The area within which Newman's text was printed measured 1/160 mm square.

5. Eigler, "From the Bottom Up." For a technical introduction to the STM, see Chen, *Introduction to Scanning Tunneling Microscopy*. Ratner and Ratner offer a brief and elementary summary of various atomic imaging techniques in *Nanotechnology*, 39–43. Wang, Lui, and Zhang provide an overview of the STM's operation and applications in *Handbook of Nanophase and Nanostructured Materials*. Perhaps the best description for the general reader is provided by Lieber, "Scanning Tunneling Microscopy." For a visual presentation of the STM's operation, see the web page of the Institute für Allgemeine Physik, http://www.iap.tuwien.ac.at/www/surface/STM_Gallery/stm_schematic.html (accessed April 28, 2016).

6. Eigler and Schweizer, "Positioning Single Atoms with a Scanning Tunneling

Microscope," 526, 524. Cf. Foster, Frommer, and Arnett, "Molecular Manipulation Using a Scanning Tunneling Microscope."

7. Dickinson, Poem #706, *The Poems of Emily Dickinson*, 314.

8. Jakobsen, *Selected Writings*, 132.

9. Benveniste, *Problems in General Linguistics*, 219.

10. Hegel, *Phenomenology of Spirit*, 60.

11. Barthes, *The Rustle of Language*, 51. For a useful summary of theories of deixis, to which I am indebted here, see Rasula, "The Poetics of Embodiment," 62–78.

12. Benveniste, *Problems in General Linguistics*, 224, 225.

13. McCaffery, *Shifters* (1976), in *Seven Pages Missing: Volume One*, 127.

14. McCaffery, "*Shifters*: A Note" (1976), in *Seven Pages Missing: Volume One*, 450.

15. Leon Rodiez qtd. in Kristeva, *Revolution in Poetic Language*, 256.

16. On the function and semiotic value of the dash in Dickinson's poetry, see, in particular, Hartman, *Criticism in the Wilderness*, 126; Howe, *My Emily Dickinson*, 23; and Miller, *Emily Dickinson*, 53. For detailed readings of "I cannot live with You –," see Cameron, *Choosing Not Choosing*; Farr, *The Passion of Emily Dickinson*; Hawkins, "Constructing and Residing in the Paradox of Dickinson's Prismatic Space"; Jackson, *Dickinson's Misery*, 142–58; Juhasz, *The Undiscovered Continent*; Kher, *The Landscape of Absence*; Kohn, "'I cannot live with You –'," 154–55; Stonum, *The Dickinson Sublime*; and Wolff, *Emily Dickinson*. As far as I have been able to discern, the features of the line that I have focused on here—this particular use of the dash or the role of shifters in this poem—have not been discussed in the criticism.

17. Nancy, "The Sublime Offering," 233.

INTRODUCTION: MATERIALS SCIENCE, MATERIALIST POETICS

1. Heidegger, "Poetically Man Dwells," 213.

2. See Heidegger, "The Question Concerning Technology," 10–13 and "Poetically Man Dwells," 213.

3. See Gimzewski and Welland, *Ultimate Limits of Fabrication and Measurement*.

4. On "the Pound tradition," see Perloff, *The Dance of the Intellect*; Beach, *ABC of Influence*; and Kenner, *The Pound Era*. Other essential guides, with a materialist perspective, to variations of the tradition I study here are Watten, *The Constructivist Moment*; Rasula, *This Compost*; Byrd, *The Poetics of the Common Knowledge*; Mackey, *Discrepant Engagement*; Davidson, *Ghostlier Demarcations*; DuPlessis, *Blue Studios*; and Dworkin, *Reading the Illegible*.

5. Zukofsky, "Program: 'Objectivists' 1931," 192.

6. Of course, the archive of poetic traditions and practices potentially relevant to my project vastly exceeds those I can address in this book. Luckily there are excellent studies and anthologies attentive to this expanded field of materialist poetry and poetics. For studies of visual poetry and typography during the first half of the

twentieth century, see Bohn, *The Aesthetics of Visual Poetry* and Drucker, *The Visible Word*. Drucker's *Experimental—Visual—Concrete* is a useful guide to more recent work. Mary Ellen Solt's *Concrete Poetry* is a classic anthology. See also Rasula and McCaffery, *Imagining Language* for a broader historical perspective.

7. Hill and Naylor, *Facture*.

8. McCaffery, *Prior to Meaning*, xv.

9. Ibid.

10. On the production of "biomaterials" by materials scientists, see Benyus, *Biomimicry* and Ball, "Only Natural," in *Made to Measure*.

11. Ball, *Designing the Molecular World*, 5.

12. Eigler, "From the Bottom Up," 425.

13. Ball, *Made to Measure*, 5.

14. Ozin and Arsenault, *Nanochemistry*, vii.

15. Mau et al., *Massive Change*, 141.

16. Olson, "Designing a New Material World." Mehmet Sarikaya qtd. in Mau et al., *Massive Change*, 142.

17. Pound, "A Retrospect," 4.

18. McCaffery, *Prior to Meaning*, xix.

19. Tiffany, *Toy Medium*, 9.

20. Though it is not the focal point of my critique here, I should say that I do not agree with Tiffany's conflation of materialist ontology and empiricist epistemology, particularly given that the empiricism he seems to have in mind is of a positivist variety. My own understanding of the epistemological requisites of scientific materialism is closer to that of Gaston Bachelard in such books as *Le matérialisme rationnel*, *Le rationalisme appliqué*, *The Philosophy of No*, and *The New Scientific Spirit*. In Bachelard's early work on scientific epistemology, scientific objectivity is not empirically grounded in "pictures" of matter, whether visual representations or imaginary figures, but rather produced by a dialectical relation of empiricism and rationalism whereby it is through the mathematical formalization of experimental evidence—and the experimental testing of these formalisms—that science constructs revisable theories concerning the objectivity of the real. It does not do so through direct observation. I discuss Bachelard further in Chapter 5.

21. Tiffany, *Toy Medium*, 171.

22. Ibid., 291.

23. Ibid., 3–4.

24. Ibid., 4.

25. Ball, "It's a Small World."

26. Daston and Galison, *Objectivity*, 401–2.

27. On theoretical "representing" and experimental "intervening" as complementary elements of scientific practice, see Hacking, *Representing and Intervening*.

28. Daston and Galison, *Objectivity*, 391–92.

29. Graham, "Pietà," 72.

30. See my reading of Dickinson's line in the preface.

31. Wang, *Mad Science in Imperial City*, 74.

32. Michaels, *The Shape of the Signifier*, 18.

33. See Cohen et al., *Material Events*.

34. *The Shape of the Signifier* opens with a critique of Howe, *The Birth Mark*.

35. Michaels, *The Shape of the Signifier*, 13.

36. Ibid., 6, 60.

37. Ibid., 11.

38. Deleuze, *Francis Bacon*, 81.

39. Wang, "The Politics of Error."

40. Johnson, "A Note," in *ARK* (unpaginated).

41. McCaffery, "Carnival," 444.

42. Ibid.

43. Ibid.

44. Ibid., 445.

45. Deleuze, *Cinema 2*, 29.

46. Ibid.

47. See Regis, *Nano!: Remaking the World Atom by Atom*.

48. Such a mechanist approach might be aligned, as James Gimzewski and Victoria Vesna have argued in their article "The Nanomeme Syndrome," with the assembler-based approach to nanoscale construction popularized by Eric K. Drexler and the Foresight Institute, following the inspiration of Feynman's 1959 address. In *The Engines of Creation*, Drexler proposes the construction of nanoscale robots ("assemblers") capable of directed mechanical interactions with atoms and molecules. "Because assemblers will let us place atoms in almost any reasonable arrangement," he writes, "they will let us build almost anything that the laws of nature allow to exist. In particular, they will let us build almost anything we can design—including more assemblers. The consequences of this will be profound, because our crude tools have let us explore only a small part of the range of possibilities that natural law permits. Assemblers will open a world of new technologies" (14). Relayed by Michael Crichton's novel *Prey* (2002), Drexler's descriptions of assemblers have played a major role in shaping popular conceptions of nanotechnology. The development of nanotechnology has not, thus far, borne out Drexler's vision, and in *The Limits of Fabrication* I grant little attention to his model of nanotechnology. I am concerned with existing fabrication techniques that have made concrete contributions to materials science or that have the potential to do so without any dependence on assemblers. On the importance of speculative discourse and science fiction in constructing both popular and scientific concep-

tions of nanotechnology, see Colin Milburn's important account, *Nanovision: Engineering the Future*. For an assessment and critique of Milburn's methodology, see Brown, "Next to Now."

49. See Whitesides and Grzybowski, "Self-Assembly at All Scales."

50. Ozin and Arsenault, *Nanochemistry*.

51. See Agamben, *The Open*.

52. Smalley, "Nanotechnology and the Next 50 Years."

53. Perloff, "The Oulipo Factor."

54. Wang, "The Politics of Error."

1. THE INORGANIC OPEN: NANOTECHNOLOGY AND PHYSICAL BEING

1. Thacker, *Biomedia*, 117.

2. Ibid., 21.

3. Ostman, "The Nanobiology Imperative."

4. Milburn, *Nanovision*, 162.

5. Agamben, *The Open*, 80 (cited hereafter in text as *TO*).

6. Heidegger, *Fundamental Concepts of Metaphysics*, 177 (cited hereafter in text as *FCM*).

7. Agamben, *Homo Sacer*, 11.

8. Ibid.

9. Von Uexküll's important work is now available in English translation under the title *A Foray into the Worlds of Animals and Humans*.

10. On Heidegger's reading of Rilke's poem, see also Santner, *On Creaturely Life*.

11. Rilke, "The Eighth Duino Elegy," 377.

12. Ibid., 379.

13. Agamben, *Homo Sacer*, 9.

14. See Agamben, *The Open*, 45–47; von Uexküll, *A Foray into the Worlds of Animals and Humans*, 44–52.

15. Graham Harman argues that Heidegger's distinctions between modalities of being on the basis of having or not-having "access" and on the basis of "the as-structure" fail with regard to any sort of being whatever. See Harman's engagement with the arguments of Heidegger's 1929–30 seminar in *Tool-Being*, 68–80. On the function of rocks as philosophical examples, see Hacking, *The Social Construction of What?* 186–206.

16. Rasmussen et al., "Transitions from Nonliving to Living Matter," 963.

17. On the rhetoric of "postbiological life" and of the "postvital" that accompanies nanotechnology's disintegration of the organism, see Milburn, "Nano/Splatter."

18. Taya, "Bio-Inspired Design of Intelligent Materials," 54.

19. Carbon nanotubes, currently the strongest known materials, are discussed at greater length in Chapter 3.

20. Jung et al., "Aligned Carbon Nanotube-Polymer Hybrid Architectures for Diverse Flexible Electronic Applications," 414.

21. Ibid.

22. Ibid.

23. Koyama et al., "Harnessing the Actuation Potential of Solid-State Intercalation Compounds," 498. The authors note that such materials "could enable generalized, large-scale structural actuation, future applications of which could include shape-morphing hulls and wings for air and water vehicles, robotics, and other 'smart' or adaptive structures" (492).

24. Kloeppel, "DNA-wrapped Carbon Nanotubes Serve as Sensors in Living Cells."

25. Strano et al., "Optical Detection of DNA Conformational Polymorphism on Single-Walled Carbon Nanotubes," 510.

26. See the Center for Embedded Networked Sensing homepage at http://www.cens.ucla.edu (accessed April 28, 2016).

27. Taya, "Bio-Inspired Design of Intelligent Materials," 54.

28. Milburn, "Nano/Splatter," 299. Cf. the chapter of the same title in *Nanovision*, 161–87.

29. Milburn, "Nano/Splatter," 299. This is also where my analysis of "physical being" diverges from Bill Brown's investigations of "the social afterlife of things." In "All Thumbs," his contribution to a 2004 *Critical Inquiry* symposium titled "The Future of Criticism," Brown calls for an "account of how our personifying instincts express a posthumanism of their own—not the denial of agency and autonomous subjectivity, but the extension of consciousness and agency to nonhuman objects." "I suspect that it is only such an extension," Brown writes, "that will enable us to fathom the sameness of what we too easily call the nonhuman, to think beyond the ontological divide between 'humans' and 'things'" (457). In my opinion, an anthropomorphic "extension of consciousness and agency" to things does not amount to an effort to think beyond the ontological divide Brown specifies but rather to an anthropocentric incapacity to think the *difference* of the nonhuman or the inorganic.

30. Milburn, "Nano/Splatter," 302.

31. On the relation between "bare life" and the Greek term *haplōs*, see Agamben, *Homo Sacer*, 182.

32. See, for example, the passage in Esposito, *Bíos*, in which he discusses Heidegger's distinction between stone, animal, and man (155–56). Just as in Agamben's work, the relevance of the distinction between stone and animal to the constitution of the category of life, and therefore of the biological, goes entirely uninterrogated. Indeed, throughout *Bíos* the distinction between living and non-living being—as that which constitutes the category of the biological—is never seriously examined by Esposito. Despite his querying of various aporia at the heart of *bíos*, Esposito seems to accept this distinction as a given.

33. Foucault, *Society Must Be Defended*, 254.

34. Campbell, *Improper Life*, 1–30.

35. Stiegler, *Technics and Time, 1*, 135 (cited hereafter in text as *TT*). Stiegler's account of the invention of the human relies heavily on theories of technology developed by Bertrand Gille, André Leroi-Gourhan, and Gilbert Simondon. See Gille, *Histoire des Techniques*; Leroi-Gourhan, *L'homme et la matière* and *Gesture and Speech*; and Simondon, *Du mode d'existence des objets techniques.*

36. Heidegger, *Being and Time*. On anticipation, see paragraphs 61 and 62. On falling and thrownness, see paragraph 38.

37. Heidegger qtd. in Nancy, *The Sense of the World*, 59 (cited in text hereafter as *SW*).

38. See Nancy, "Of Being Singular Plural."

39. Cf. Nancy, *Corpus*, 15–19.

40. Nancy, "Res ipsa et ultima," 316.

2. OBJECTISM: CHARLES OLSON'S POETICS OF PHYSICAL BEING

1. Olson, "Projective Verse," 247.

2. Harman, *Tool-Being*, 225.

3. Ibid., 290.

4. See Harman, "On Vicarious Causation."

5. Harman, *Tool-Being*, 287.

6. For a longer critique of "object-oriented ontology," see Brown, "The Nadir of OOO." See also Brown, "Speculation at the Crossroads," a review of Tristan Garcia's *Form and Object.*

7. The key concepts cited here are elaborated in Latour, *We Have Never Been Modern* and *Reassembling the Social.*

8. For a brief but decisive critique of the series of reductions upon which Latour's supposed commitment to "irreduction" turns, see Brassier, "Concept and Object," 51–54.

9. See Grusin, *The Nonhuman Turn.*

10. Olson, "Projective Verse," 247. A note on Olson's gendered language: though I have chosen not to insert "sic" after every universalizing usage of the term "man" in Olson's writings, I want to register my recognition that this usage is complicit with the masculinism running throughout his poetry and poetics. In Chapter 5, I take up this problem in detail by considering Caroline Bergvall's work as a feminist/queer rearticulation of objectist poetics. On the contexts and effects of Olson's masculinism, see DuPlessis, "Manifests" and "Manhood and Its Poetic Projects." See also Davidson, *Guys Like Us*, 28–48.

11. Conte, *Unending Design*, 28.

12. Perloff, *Radical Artifice*, 134–35.

13. McCaffery, *Prior to Meaning*, 48.

14. Nelson, "Organic Poetry."

15. Byrd, *The Poetics of the Common Knowledge*, 347–70. For an introduction to Maturana and Varela's theory of autopoietic systems, see Maturana and Varela, *The Tree of Knowledge*. For their more technical presentation, see *Autopoiesis and Cognition*. I return to the relation of objectism to autopoietic theory later in this chapter.

16. Altieri, *Enlarging the Temple*, 102.

17. Olson, *Proprioception*, 181.

18. Olson, *Letters for Origin*, 5 (italics in original).

19. Olson, "Projective Verse," 243.

20. Ibid., 247.

21. Bergson, *Creative Evolution*, 126 (italics in original).

22. Bennett, *Vibrant Matter*. Bennett writes: "maybe it's worth running the risks associated with anthropomorphizing (superstition, the divinization of nature, romanticism) because it, oddly enough, works against anthropocentrism: a chord is struck between person and thing, and I am no longer above or outside a nonhuman 'environment'" (120). I disagree with the notion that the projection of human attributes into things functions as a bulwark against anthropocentrism. I would argue instead that anthropocentrism is only reinforced by an incapacity to consider that which is indeed nonhuman *as* nonhuman—but not therefore necessarily inert or passive.

23. Jones, *Soft Machines*, 13.

24. On the role of quantum mechanics in nanoscale engineering, see Milburn, *Schrödinger's Machines*.

25. Ratner and Ratner, *Nanotechnology*, 7.

26. Ibid.

27. Banerjee qtd. in Hsu, "Working Toward 'Smart Windows.'" See Banerjee, Whittaker, and Patridge, "Microscopic and Nanoscale Perspective of the Metal Insulator Phase Transitions of VO2."

28. Olson, "Equal, That Is, to the Real Itself," 121.

29. Ibid.

30. Olson, *The Special View of History*, 29.

31. Ibid., 58.

32. Olson's debt to Whitehead has been well documented. My specific goal here is to address Whitehead's influence with regard to "objectism," insofar as the latter involves a theory of physical being that distinguishes Olson's poetics from conceptions of "organic form" with which his work is often linked. Accounts of the relation between Olson and Whitehead that I have found useful are Blaser, "The Violets"; Byrd, *Charles Olson's Maximus*; and Hoeynck, *The Principle of Measure in Composition by Field*. I find Robert von Hallberg's treatment of the relation between Olson and Whitehead in *Charles Olson* deeply flawed. Von Hallberg fails to recognize the

ontological, rather than epistemological, significance of Olson's "stance toward reality," an error at the root of his argument that Olson reintroduces a humanist standpoint into Whitehead's cosmology.

33. Whitehead, *Process and Reality*, xiv (cited hereafter in text as *PR*).

34. Whitehead, *Science and the Modern World*, 49 (cited hereafter in text as *SMW*).

35. Whitehead, *The Concept of Nature*, 145. In the vocabulary of these 1920 lectures, the term "ingression" denotes "the general relation of objects to events." According to Whitehead, "The ingression of an object into an event is the way the character of the event shapes itself in virtue of the being of the object. Namely the event is what it is, because the object is what it is; and when I am thinking of this modification of the event by the object, I call the relation between the two 'the ingression of the object into the event.' It is equally true to say that objects are what they are because events are what they are. Nature is such that there can be no events and no objects without the ingression of objects into events" (144).

36. A "prehension," in Whitehead's vocabulary, is a determinate relation. It is the manner in which one actual entity or actual occasion selectively includes another within its own constitution. A "negative prehension" is the definite exclusion of one actual entity from positive contribution to another's internal constitution (*Process and Reality*, 41).

37. Perloff, *Radical Artifice*, 134–35.

38. Charles Olson to Robert Creeley, April 7, 1951, in Olson, *Charles Olson & Robert Creeley*, vol. 5, 127–28.

39. Olson, "Equal, That Is, to the Real Itself," 125.

40. Ibid.

41. Ibid.

42. Ibid., 121.

43. Pound, *The Cantos of Ezra Pound*, LXXXI, 538.

44. See Finch, *The Ghost of Meter*.

45. Levertov, "Some Notes on Organic Form," 67–68.

46. Duncan, "A Poem Beginning with a Line by Pindar," 69.

47. Ibid., 64.

48. Duncan, "The Structure of Rime I," 12.

49. Olson, "Projective Verse," 240.

50. Levertov, "Some Notes on Organic Form," 73 (italics in original).

51. See Kenner, *The Pound Era*.

52. Pound, *The Cantos of Ezra Pound*, LXXXI, 541.

53. Ibid., XCIII, 648.

54. Ibid., LXXXI, 541.

55. Ibid., 822.

56. Ibid., LXXXI, 541.

57. Ibid., XC, 625.

58. Olson, "Projective Verse," 240.

59. Olson, *The Maximus Poems*, I.52 (cited hereafter in text as *M*).

60. Ibid., I.14–15.

61. Robert Creeley to Cid Corman, September 30, 1950, in Corman, *Origin* 2, 92.

62. Ibid.

63. Olson, "Human Universe," 161.

64. Olson, "Projective Verse," 242.

65. McCaffery, *Prior to Meaning*, 46.

66. Olson, "Projective Verse," 244.

67. Ibid., 244.

68. Whitehead, *Process and Reality*, 48.

69. Olson, "Projective Verse," 243–44.

70. Olson, *Proprioception*, 181.

71. Ibid.

72. Olson, "Human Universe," 161.

73. Olson, "Projective Verse," 248.

74. On the Scanning Tunneling Microscope, see the prologue.

75. Olson, *The Special View of History*, 49.

76. Pynchon, *Gravity's Rainbow*, 712.

77. Olson, "Projective Verse," 247.

78. Dworkin, *Strand*, 9; Bök, *Crystallography*, 156; Lin, *BlipSoak01*, 13; Bergvall, *Fig*, 33.

79. Olson, "Projective Verse," 247.

80. Altieri, *Enlarging the Temple*, 98–99.

81. Perloff, "Charles Olson and the 'Inferior Predecessors,'" 295.

82. Von Hallberg, *Charles Olson*, 84.

83. Ibid., 113.

84. Note that Whitehead is not referring to objectivist poetics but to objectivism as a philosophical position. My argument here is that objectivist poetics is more or less consistent with the epistemological position attributed to philosophical objectivism by Whitehead.

85. Zukofsky, "An Objective," 16.

86. Oppen, "Route," 293.

87. Oppen, *Selected Letters of George Oppen*, 300.

88. Oppen, "Route," 193.

89. Oppen, *Of Being Numerous*, 186.

90. Zukofsky, "A," 563.

91. On Pound's complex figuration of the poetic image, see Tiffany, *Radio Corpse*.

92. Albright, *Quantum Poetics.*

93. Olson, "Projective Verse," 247.

94. See Maturana and Varela, *Autopoiesis and Cognition.* In *How We Became Posthuman*, Hayles offers a critical assessment of autopoietic theory and charts the widening divide between Maturana's and Varela's perspectives during the 1980s and 1990s. Varela's later work is developed in *The Embodied Mind*, coauthored with Evan Thompson and Eleanor Rusch. For recent essays on the import of autopoietic theory and Luhmann's system theory, see Clarke and Hansen, *Emergence and Embodiment.*

95. Luhmann's major work is *Social Systems.*

96. Maturana and Varela, *Autopoiesis and Cognition*, 81.

97. Wiener, *Human Use of Human Beings*, 28.

98. Maturana and Varela, *Autopoiesis and Cognition*, 80.

99. Byrd, *Poetics of the Common Knowledge*, 291.

100. Ibid., 349.

101. Ibid., 294.

102. Olson, "Projective Verse," 247.

103. Maturana and Varela, *Autopoiesis and Cognition*, 127.

104. Olson, "Projective Verse," 247.

105. Maturana and Varela, *Autopoiesis and Cognition*, 73.

106. Nancy, *The Sense of the World*, 58.

107. Ibid., 63.

108. Whitehead, *The Concept of Nature*, 29, 41.

109. Pound, *Cantos*, CXVI, 816.

110. Olson, "In Cold Hell, In Thicket," 29.

111. Ibid., 32.

112. Olson, "The Kingfishers," 9.

3. DESIGN SCIENCE: GEODESIC ARCHITECTURE IN NANOSCALE CARBON CHEMISTRY AND RONALD JOHNSON'S *ARK*

1. Johnson, "A Note," in *ARK. ARK* is unpaginated. References to the poems will be made in text hereafter, by section number.

2. Perloff, "Songs of the Earth," 204.

3. Kroto, "Macro-, Micro-, and Nano-scale Engineering," 234.

4. Ibid.

5. Ibid., 232.

6. Johnson, "A Note on ARK," in *ARK 50: Spires 34–50*, 56–57. Johnson composed *RADI OS* (1977) by excising words from the first four books of *Paradise Lost* while leaving the remaining text in its position on the page. He thus creates what he calls a "ventilated" version of Milton's work, which visually resembles an open field poem.

7. For various accounts of this event, see Katz, *Black Mountain College*, 148–49;

Marks and Fuller, *Dymaxion World*, 182; Krausse and Lichtenstein, *Your Own Private Sky*, 318; Sieden, *Buckminster Fuller's Universe*, 312–13; and Kenner, *Bucky*, 234–35.

8. Elaine de Kooning qtd. in Harris, *The Arts at Black Mountain College*, 151.

9. John Cage qtd. in Krausse and Lichtenstein, *Your Own Private Sky*, 320.

10. Marks and Fuller, *Dymaxion World*, 182.

11. Merce Cunningham offers this account during an interview for the documentary *Thinking Out Loud: Buckminster Fuller* (1995), dir. Karen Goodman and Kirk Simon, qtd. in Krausse and Lichtenstein, *Your Own Private Sky*, 318.

12. Fuller, *Synergetics*, 3.

13. Fuller, *Critical Path*, 232.

14. Fuller, *Synergetics*, 6–7.

15. Fuller, *Critical Path*, 158.

16. Whitehead, *Process and Reality*, 18.

17. Ibid.

18. Olson, "Projective Verse," 243.

19. Robert Creeley to Cid Corman, September 30, 1950, in Corman, *Origin* 2, 92.

20. Fuller, *Synergetics*, 14–15.

21. Ibid., 108.

22. Fuller qtd. in Krausse and Lichtenstein, *Your Own Private Sky*, 278.

23. Fuller, *Synergetics*, 108.

24. Charles Olson to Robert Creeley, May 15, 1952, in Olson, *Charles Olson & Robert Creeley*, vol. 10, 68.

25. For more detailed accounts of Fuller's work on energetic geometry during these years, see Fuller and Marks, *Dymaxion World*, 39–49; Krausse and Lichtenstein, *Your Own Private Sky*, 276–313; and Kenner, *Bucky*, 105–18, 234–41. For a complete elaboration of Fuller's experiments with the closest packing of spheres, see *Synergetics*, 108–35; on the division of the sphere into great circles, see 164–89.

26. Fuller and Marks, *Dymaxion World*, 43.

27. Kenner, *Bucky*, 236–37. For other descriptions of the same dome, see Fuller and Marks, *Dymaxion World*, 183–84; Sieden, *Buckminster Fuller's Universe*, 314–15; and Krausse and Lichtenstein, *Your Own Private Sky*, 326.

28. Fuller, *Inventions*, 179.

29. Ibid.

30. Olson, "Projective Verse," 243 (my italics).

31. Pynchon, *Gravity's Rainbow*, 411–12.

32. For an English translation of the 1890 speech in which Kekulé related his dream, see Benfey, "August Kekulé and the Birth of the Structural Theory of Organic Chemistry in 1858." As Philip Ball notes, another German chemist, Johann Loschmidt, had in fact reported the cyclical structure of benzene four years prior to Kekulé's speech. See Ball, *Designing the Molecular World*, 31. See also the skeptical account of

Kekulé's dream offered by Bensaude-Vincent and Stengers in *A History of Chemistry*, 155.

33. Kekulé, "Sur la constitution des substances aromatiques."

34. Baggott, *Perfect Symmetry*, 57. See also Ball, *Designing the Molecular World*, 42; and Applewhite, "The Naming of the Buckminsterfullerene," 332–36.

35. Kemp, "Kroto and Charisma." "A crucial component that Kroto brought to the buckyball team," writes Kemp, "was his natural instinct as a designer. Indeed, he had long hankered after a career in graphic design, and, before the consuming success of C60, planned to found a studio for scientific graphics. He is one of those scientists, like Leonardo and Kepler, naturally drawn to the tangible beauty of complex symmetries in works of nature and art."

36. Kemp, "Kroto and Charisma."

37. Thompson, *On Growth and Form*, 157–58 (italics in original). On the relation of Euler's Formula, the geometry of polyhedral forms, and Fuller's geodesic design principles to the modeling of C60, see Baggott, *Perfect Symmetry*, 79–92.

38. Baggott *Perfect Symmetry*, 58; Ball, *Designing the Molecular World*, 44.

39. Baggott, *Perfect Symmetry*, 68–69; Ball, *Designing the Molecular World*, 44.

40. Kroto et al., "C60." Kroto defends the appropriateness of the molecule's name in "The Stability of the Fullerenes C*n*, with n=24, 28, 32, 36, 50, 60 and 70."

41. On the significance of this award for the history of carbon chemistry, see Ball, "Fullerenes Finally Score as Nobel Committee Honors Chemists."

42. Iijima, "Helical Microtubules of Graphitic Carbon."

43. Ibid., 56.

44. Ebbesen and Ajayan, "Large-Scale Synthesis of Carbon Nanotubes." See also the commentary on this essay by Dresselhaus, "Down the Straight and Narrow."

45. Kroto, "Macro-, Micro-, and Nano-Scale Engineering," 234. Nanotechnologists call sheets of nanotubes "bucky paper" and the secondary impurities that accrete to nanotubes during fabrication processes are referred to as "bucky goo." See Baughman, Zakhidov, and de Heer, "Carbon Nanotubes."

46. *AZoNano*, "Nanotube Production Capacities and Demand." On the impact of nanotubes on carbon chemistry from 1991 to 2001, see Ball, "Roll Up for the Revolution." For a technical but clear introduction to carbon nanotube chemistry, see Harris, *Carbon Nanotubes and Related Structures*.

47. Ajayan and Tour, "Nanotube Composites," 1066.

48. Ibid., 1067. On the mechanics of nanotubes and the intermolecular interactions operative between them, see Yakobseon and Couchman, "Carbon Nanotubes," 587–601.

49. Kroto, "Macro-, Micro-, and Nano-Scale Engineering," 234.

50. See Smalley, "Future Global Energy Prosperity." On Fuller's proposal for an "omni-world-integrating electrical-energy network grid," see *Critical Path*, 252–69.

51. Smalley qtd. in Ball, *Designing the Molecular World*, 52. See also Applewhite, "The Naming of the Buckminsterfullerene," 332–36.

52. Kroto, "Macro-, Micro-, and Nano-Scale Engineering," 232.

53. Ozin and Arsenault, *Nanochemistry*, 509.

54. Applewhite, "The Naming of the Buckminsterfullerene," 334.

55. Ibid.

56. Ozin and Arsenault, *Nanochemistry*, 553.

57. Whitesides and Grzybowski, "Self-Assembly at All Scales," 2421 (italics in original).

58. Ibid.

59. Mau et al., *Massive Change*, 16.

60. Ibid., 141.

61. Ozin and Arsenault, *Nanochemistry*, 510.

62. See Milburn, *Nanovision*.

63. I refer to the title of Vendler's book, *Part of Nature, Part of Us*.

64. I refer to the title of Ball's study of materials research and engineering, *Designing the Molecular World*.

65. Mau et al., *Massive Change*, 140.

66. Johnson, "From the Notebooks," 76.

67. Ibid., 70.

68. Ibid., 69–70.

69. Asked about the title of his poem, Johnson said, "I just thought *ARK* because it also included the rainbow—the rainbow goes all the way through the poem, that kind of arch, 'arc.'" See Johnson, "An Interview with Ronald Johnson," 34.

70. The line "remove above" may refer to Fuller's polemic against the terms "up" and "down," which, he argued, were conceptual residues of flat-earth cartography and cosmology. Fuller suggested they be expelled from the lexicon and that their denotative functions be subsumed by "out" and "in." Cf. Fuller, *Critical Path*, 54–55, 253–54.

71. Johnson quotes the passage from James's *The Coxon Fund* in which the narrator describes the "splendid intellect" of Frank Saltram.

72. Johnson, "An Interview with Ronald Johnson," 39.

73. Ibid.

74. Scroggins, "'A' to *ARK*," 146.

75. Ibid., 147.

76. Johnson, "A Note," in *ARK*.

77. Johnson, "An Interview with Ronald Johnson," 35.

78. Ibid.

79. Johnson erroneously attributes to Olson Pound's definition of the epic as "a poem including history."

80. Johnson, "An Interview with Ronald Johnson," 33.

81. Johnson, "Planting the Rod of Aaron," 2.

82. Johnson, "An Interview with Ronald Johnson," 37.

83. Elaine de Kooning qtd. in Harris, *The Arts at Black Mountain College*, 151.

84. Johnson "An Interview with Ronald Johnson," 37.

85. See, for example, the section of Fuller's *Synergetics* titled "Basic Disequilibrium LCD Triangle": "The basic right triangle as the lowest common denominator of a sphere's surface includes all the data for the entire sphere, and is the basis of all geodesic dome calculations" (480–81).

86. Fuller qtd. in Applewhite, "The Naming of the Buckminsterfullerene," 336.

87. Johnson, "Hurrah for Euphony," 29.

88. Scroggins, "'A' to *ARK*," 149.

89. Davenport, introduction to Johnson, *Valley of the Many-Colored Grasses*, 9.

90. Fuller, *Untitled Epic Poem on the History of Industrialization*; Fuller, *No More Secondhand God and Other Writings*. Fuller's third "poem," *Intuition*, was published subsequent to Davenport's remark.

91. Davenport, introduction to Johnson, *Valley of the Many-Colored Grasses*, 9–10.

92. Johnson, *Valley of the Many-Colored Grasses*, 49–50.

93. Fuller, *Synergetics*, 15.

94. Johnson, "An Interview with Ronald Johnson," 44.

95. Johnson, *The Book of the Green Man*, 11.

96. Johnson, *Valley of the Many-Colored Grasses*, 84–88.

97. On Johnson's transitional verbivocovisual work in *Songs of the Earth*, see Perloff, "Songs of the Earth."

98. Johnson, "The Unfoldings," in *Valley of the Many-Colored Grasses*, 111.

99. Johnson, "Hurrah for Euphony," 29.

100. DuPlessis, "Echological Scales," 113.

101. Albright, *Quantum Poetics*.

102. On quantum corrals, see the two classic articles by Crommie, Lutz, and Eigler: "Imaging Standing Waves in a Two-Dimensional Electron Gas" and "Confinement of Electrons to Quantum Corrals on a Metal Surface." See also Heller et al., "Scattering and Absorption of Surface Electron Waves in Quantum Corrals"; and Ball, "Round 'Em Up."

103. "What is popularly called Transcendentalism among us, is Idealism; Idealism as it appears in 1842," Emerson writes in "The Transcendentalist," 81. "The idealist," he says, "in speaking of events, sees them as spirits. He does not deny the sensuous fact: by no means; but he will not see that alone. He does not deny the presence of this table, this chair, and the walls of his room, but he looks at these things as the reverse side of the tapestry, as the *other end*, each

being a sequel or completion of a spiritual fact which more nearly concerns him" (81–82).

104. Olson, "Projective Verse," 157.

105. Ibid., 247.

106. Emerson, "Nature," 3.

107. Olson, "Human Universe," 157.

108. Olson, "Against Wisdom as Such," 263.

109. Rasula, *This Compost*, 168. Whereas Rasula situates Olson's work along a continuum running from romanticism through transcendentalism and Poundian modernism, I would mark the distance articulated here between Olson and Emerson as the crucial distinction between objectism and the idealism endemic to models of organic form.

110. I discuss the pertinence of Gilles Deleuze's concept of "signaletic material" to materialist poetics in the introduction. See Deleuze, *Cinema 2*, 29.

111. Johnson, "A Note," in *ARK*.

112. Johnson, "Hurrah for Euphony," 25.

113. Kenner, *The Pound Era*, 146.

114. Ibid., 150.

115. Ibid., 161–62.

116. Ibid., 157. Kenner quotes from Emerson, "The Poet." Cf. Emerson, *Essential Writings*, 292.

117. Kenner's *Geodesic Math and How to Use It* was the most requested backlist title of the University of California Press after going out of print in 1990, and it has since been reissued in paperback (Berkley: University of California Press, 2003).

118. Kenner, *Bucky*, 29.

119. Ibid., 43.

120. Ibid., 295.

121. DuPlessis, "Echological Scales," 103.

122. Johnson, *Valley of the Many-Colored Grasses*, 25.

123. Johnson, "An Interview with Ronald Johnson," 44.

124. Ibid.

125. A volume titled *The Shrubberies* was edited by Peter O'Leary and published with Flood Editions in 2001. In the afterword, O'Leary states that an edition of *The Outworks* is forthcoming with the same press, but thus far no such volume has been published. "Architecture has a plan and an execution and a dedication ceremony," but only, Johnson jokingly concedes, "if they can ever get their people together." Johnson, "An Interview with Ronald Johnson," 35.

126. Fuller, *Synergetics*, 17.

127. Levertov, "Some Notes on Organic Form," 68.

128. By this standard, a poem like Gary Snyder's *Rivers and Mountains Without*

End, with its contingent structure of inscribed encounters and its ceaseless interrogation of irreducibly *contextual* ecologies, is less a "nature poem" than *ARK*. Compare the punctual decompletion of conclusion in Snyder's final lines, "The space goes on. / But the wet black brush / tip drawn to a point, / lifts away," to Johnson's "Omphalos triumphant" and "music of spheres solved." *Rivers and Mountains* does not propose, like *ARK*, the construction of "a monument" ("An Ozimandias of the spirit," as Johnson says) that would instantiate a universal order but rather the punctual inscription of irreducibly specific situations and environments, neither immersed nor included in any higher order.

129. Fuller, "Conceptuality of Fundamental Structures," 66. On Fuller's aversion to irrational numbers, see Kenner, *Bucky*, 132–39.

130. Fuller, *Synergetics*, 23.

131. Whitehead, *Process and Reality*, 214–15.

132. See Badiou, *Being and Event*, 123–70.

4. SURRATIONAL SOLIDS, SURREALIST LIQUIDS: CRYSTALLOGRAPHY AND BIOTECHNOLOGY IN MATERIALS SCIENCE AND MATERIALIST POETRY

1. Cairns-Smith, "The Origin of Life and the Nature of the Primitive Gene."

2. See also Cairns-Smith, *Life's Puzzle*; Cairns-Smith, *Seven Clues to the Origin of Life*; and Cairns-Smith and Harman, *Clay Minerals and the Origin of Life*.

3. Cairns-Smith, *Genetic Takeover*, 371.

4. Cairns-Smith's theory is discussed by Dawkins in *The Blind Watchmaker*, 148–58; by Dennett in *Darwin's Dangerous Idea*, 157–58; and by Ball in *Designing the Molecular World*, 265–66. His theory has also inspired more recent experimental work by researchers in Caltech's DNA and Natural Algorithms Group. See Schulman and Winfree, "How Crystals That Sense and Respond to Their Environments Could Evolve" and "Self-Replication and Evolution of DNA Crystals."

5. For an introduction to the techniques of modern crystallography, see Glusker and Trueblood, *Crystal Structure Analysis*. On the application of X-ray crystallography to molecular biology, see Rodes, *Crystallography Made Crystal Clear*. Rodes's text is particularly notable for its focus on the epistemological/methodological problem of the relation between "model" and "structure."

6. On "the unacknowledged exploitation of Rosalind Franklin's x-ray crystallographic data" by Watson and Crick, see Magner, *A History of the Life Sciences*, 459. For a biography of Franklin, see Maddox, *Rosalind Franklin*. Lee discusses the contribution of X-ray crystallography to molecular biology, situating this issue in relation to philosophical debates concerning the distinction between artifacts and organisms, in *Philosophy and Revolutions in Genetics*, 119–29.

7. RCSB Protein Data Bank, http://www.rcsb.org/pdb/statistics/holdings.do.

8. "Whither Crystallography."

9. Lieber, "Scanning Tunneling Microscopy," 28. Leiber's reference to "condensed matter research" alludes to the field of condensed matter physics (CMP), the study of the physical properties of solids and liquids from the perspective of quantum and statistical mechanics. It is the field of physics most closely aligned with materials research, and it overlaps broadly with nanotechnology. It also increasingly overlaps with molecular biology and "biological physics." The following description of CMP's capacious purview, from the Science and Engineering web pages at the University of Edinburgh, nicely captures the contemporary intersection of solid-state chemistry and molecular biology: "Because the subject matter of CMP is the very stuff of the everyday world, it has very wide applicability. Thus, CMP overlaps considerably with materials science. . . . Earth scientists find that results from CMP can help them in understanding mineralogical phase transformations. Increasingly, condensed matter physicists are unraveling the complex behavior of suspensions and polymers, which are important in everything from paints through shampoo to tomato ketchup. Finally, the methods and results of CMP are now being applied to the study of biological systems, which are, after all, specialized forms of condensed matter." http://www.ph.ed.ac.uk/cmatter/.

10. Lieber, "Scanning Tunneling Microscopy," 39.

11. See "Layer-by-Layer Self-Assembly," in Ozin and Arsenault, *Nanochemistry*, 95–127.

12. On materials science applications involving zeolites and porous solids, see Ozin and Arsenault, *Nanochemistry*, 11–14 as well as Ball, *Made to Measure*, 282–312.

13. Goodsell, *Bionanotechnology*, 8.

14. Ibid., 5.

15. Ibid., 92.

16. For a popular introduction to biomineralization, see Benyus, *Biomimicry*, 95–145. On efforts to replicate the production of abalone shell in particular, see 98–111. Ball also offers a helpful account of biomineralization in *Made to Measure*, 192–204. For a more technical introduction, see the special issue of *Reviews in Mineralogy & Geochemistry* titled "Biomineralization," ed. Dove, De Yoreo, and Weiner. In particular, see Steve Weiner and Patricia M. Dove, "An Overview of Biomineralization Processes and the Problem of the Vital Effect," 1–30; John S. Evans, "Principles of Molecular Biology and Biomacromolecular Chemistry," 31–56; and James J. De Yoreo and Peter G. Vekilov, "Principles of Crystal Nucleation and Growth," 57–90.

17. Benyus, *Biomimicry*, 106.

18. Ibid., 107.

19. Ibid., 108.

20. Ibid., 118.

21. Smalley, "Nanotechnology and the Next 50 Years."

22. Bök, *Crystallography*, 156. I cite here from the revised edition of Bök's volume. *Crystallography* was originally published in 1994 and then revised and reissued (in order to improve the text's layout) in 2003. Cited in text hereafter as *C*.

23. Bergvall, *Goan Atom*, 7 (cited in text hereafter as *GA*).

24. Milne, "A Veritable Dollmine."

25. Dewdney, "Parasite Maintenance," 19.

26. Ibid., 23.

27. Ibid., 25.

28. Ibid., 19.

29. Ibid., 21.

30. See Baudrillard, *The Ecstasy of Communication*, 77–95. As my epigraph to the present chapter makes clear, Bök draws upon Baudrillard's notion of "the revenge of the object," figured by the crystal, in his *'Pataphysics*.

31. Bök, *'Pataphysics*, 89. Here Bök is explicating bpNichol's use of this device in order to "[prove] mathematically that 'faith' does indeed equate with 'hope.'" As Bök notes, Nichol, in turn, borrows the technique from the OuLiPo.

32. I use the general term "homograph" rather than the more specific "homonym" because the former term is also used in geometry (referring to a relation of correspondence between the points of one figure to those of another). This geometrical sense of the term is activated by the visual pun in Bök's poem, which relies upon the correspondence of points on a plane section to the points of a three dimensional object. Bök thus plays on the relation between the linguistic and geometrical homography of "diamond." Dewdney's 'pataphysical theory of autonomous linguistic sentience also contains a discussion of "Homographs and the Discharde [*sic*] of Connotation in the Poem." See Dewdney, "Parasite Maintenance," 34–35.

33. Dworkin, "Shift," in *Strand*.

34. McCaffery and Nichol, *Rational Geomancy*.

35. Olson, *Charles Olson in Connecticut* (unpaginated).

36. Olson, *Maximus Poems*, II.9, II.10.

37. McCaffery, "Carnival," 446–47.

38. Dewdney, *A Paleozoic Geology of London, Ontario* (unpaginated).

39. McCaffery, "Strata and Strategy," 191.

40. Olson, *Maximus Poems*, II.10.

41. Ibid.

42. Olson, "Projective Verse," 240.

43. Ibid., 243–44.

44. Ibid., 240.

45. Ozin and Arsenault, *Nanochemistry*, xxix–xxx (italics in original).

46. Bök, *'Pataphysics*, 5.

47. See Rancière, *The Politics of Aesthetics.*

48. Thacker, *Biomedia*, 132.

49. Doyle, *On Beyond Living*, 8. On the rhetorical impasses of such terms as "postvital" and "postbiological," see my critique in Chapter 1.

50. Thacker, *Biomedia*, 121.

51. Goodsell, *Bionanotechnology*, 12.

52. Ibid., 69.

53. Whitehead, *Process and Reality*, 214–15.

54. Olson, *Proprioception*, 181.

55. See, in particular, DuPlessis, "Manhood and Its Poetic Projects" and "Manifests." See also Davidson, *Guys Like Us*; and Mossen, "'In Thicket.'"

56. Olson, "Projective Verse," 247 (italics added).

57. DuPlessis, "Manhood and Its Poetic Projects."

58. Ibid.

59. Ibid.

60. Butler, *Bodies That Matter*, 250.

61. Ibid.

62. Ibid.

63. Butler, *Gender Trouble*, 22.

64. Audio files of Olson and Bergvall reading the following two passages can be accessed at *PennSound*: https://media.sas.upenn.edu/pennsound/authors/Olson/Boston-62/Olson-Charles_18_In-Cold-Hell_Boston_06-62.mp3; https://media.sas.upenn.edu/pennsound/authors/Bergvall/Studio-111/Bergvall-Caroline_from-Goan-Atom-1_01_UPenn_4-6-05.mp3 (accessed April 28, 2016).

65. Olson, "In Cold Hell, In Thicket," 29.

66. On Bellmer's work, see Taylor, *Hans Bellmer*. On the relation of Bellmer's dolls to materialist poetics, see Tiffany, *Toy Medium*, 87–94.

67. Bergvall, "Speaking in Tongues."

68. Thacker, *The Global Genome*, 45, 44.

69. Sedgwick, *The Epistemology of the Closet*, 28.

70. In her interview with Stammers, Bergvall notes, "My motivation has been very much to do with gender and very much to do with sexuality. These are very strong motivators which to me are to do with how you use language to construct or de-structure assumptions about gender, about sexuality, about female gender. Where do you situate the use of language within that so that you don't fall into a kind of identity-based writing, or identity-based art, but so that the whole question of identity becomes questioned. . . . There are other writers such as Wittig and Brossard, who are two conceptual French lesbian writers, or Kathy Acker or Dennis Cooper, who have developed ways of trying to deal with language in a conceptual manner so that

they could find a language that might bring those aspects of the body and of sexualized, unstraightened bodies into language" (*Speaking in Tongues*).

71. Kristeva, *Revolution in Poetic Language*. See, especially, 25–30.

72. Haraway, *Modest_Witness@Second_Millennium.FemaleMan©_Meets_OncoMouse™*, 119–72.

73. Gene Ontology Annotation Database, http://www.ebi.ac.uk/GOA/newto.html (accessed January 13, 2008).

74. Drexler, "Machines of Inner Space," 344–45, 325.

75. An audio file of Bergvall reading *Ambient Fish* is available at *PennSound*: http://media.sas.upenn.edu/pennsound/authors/Bergvall/Bergvall-Caroline_02_Ambient-Fish_Frequency_2004.mp3> (accessed January 13, 2008).

76. Bellmer, "Memories of the Doll Theme," 33. Cf. Bergvall, *Goan Atom*, 21.

77. Thacker, *The Global Genome*, 123.

5. THE SCALE OF A WOUND: NANOTECHNOLOGY AND THE POETICS OF REAL ABSTRACTION IN SHANXING WANG'S *MAD SCIENCE IN IMPERIAL CITY*

1. Wang, *Mad Science in Imperial City*, 74 (cited hereafter in text as *MS*).

2. Wang, "The Politics of Error."

3. Stefans, "Make It New."

4. Wang, "The Politics of Error."

5. Ibid.

6. Ibid.

7. See Krausse and Lichtenstein, *Your Own Private Sky*, 596. On Fuller's "World Game" as a practical instance of this imperative that drew together game theory and systems theory, see 464–97, where the influence of Fuller's work on *The Whole Earth Catalog* is also discussed.

8. Martin, *The Organizational Complex*, 40.

9. Mandel, *Late Capitalism*, 55–65.

10. See ibid., 63. Mandel cites statistics from Bairoch, *Diagnostic de l'evolution economique du Tiers-Monde*, 165.

11. On cognitive mapping, see Jameson, *Postmodernism*, 297–418.

12. Watten, "Radical Particularity, Critical Regionalism, and the Resistance to Globalization."

13. Clover, "Interview with Joshua Clover."

14. The prefix "nano" is employed seventeen times in Wang, *Mad Science in Imperial City*.

15. *Oxford English Dictionary*, "Fabrication."

16. Sohn-Rethel, *Intellectual and Manual Labor*, 48–49.

17. Ibid., 56.

18. Ibid., 57.

19. I use the term "real subsumption" in the technical sense defined by Marx in "Results of the Immediate Process of Production" to denote a historical transition from the predominance of absolute surplus value extraction (formal subsumption) to the predominance of relative surplus value extraction (real subsumption), a transition involving tendentially increased investment in constant capital relative to variable capital. See Marx, *Capital Volume 1*, 1019–25.

20. Debord, *The Society of the Spectacle*, 31.

21. "The concept of immaterial labor was filled with ambiguities. Shortly after writing those articles I decided to abandon the idea and haven't used it since. One of the ambiguities it created had to do with the concept of immateriality. Distinguishing between the material and the immaterial was a theoretical complication we were never able to resolve." Lazzarato qtd. in Iles and Vishmidt, "Work, Work Your Thoughts, and Therein See a Siege," 137–38.

22. Lazzarato, "Immaterial Labor," 133–34.

23. For a useful empirical and theoretical analysis of de-industrialization and its consequences, see Benanav and *Endnotes*, "Misery and Debt." On the basis of statistical analysis, Benanav notes that "de-industrialization is not caused by the industrialization of the 'third world.' Most of the world's industrial working-class now lives outside the 'first world,' but so does most of the world's population. The low-GDP countries have absolutely more workers in industry, but not relative to their populations" (41).

24. Boltanski and Chiapello, *The New Spirit of Capitalism*, 75.

25. Ibid., 112.

26. Virno, *A Grammar of the Multitude*, 52–63.

27. Boltanski and Chiapello, *The New Spirit of Capitalism*, 109.

28. Malabou, *What Should We Do with Our Brain*, 40–46. On Malabou's development of the concept of "plasticity," see *Plasticity at the Dusk of Writing*.

29. Malabou, *What Should We Do With Our Brain*, 5.

30. Ibid., 9.

31. Pound, "A Retrospect," 4–5.

32. Hui, "The 1989 Social Movement and the Historical Roots of China's Neoliberalism," 57. Hui elaborates: "I make this distinction (that is, between the old state and the state promoting reform) not to deny the connections between the two, but rather to point out how social conditions and the functions of the state had been transformed" (57).

33. On this point, see Slavoj Žižek's commentary on the relation between Sohn-Rethel and Louis Althusser, *The Sublime Object of Ideology*, 16–21.

34. See Derrida, "Signature Event Context."

35. Lacan, *Seminar XX*, 94.

36. Deleuze, *Difference and Repetition*, 176; Deleuze, *Différence et Répétition*, 182.

37. Deleuze, *Difference and Repetition*, 176.

38. Ibid., 181. Paul Patton translates the French *insensible* as "imperceptible." I have altered his translation by adopting the English cognate instead, since the term "insensible" retains a crucial resonance with sensibility and with sense (*sens*).

39. See Deleuze, "Ideas and the Synthesis of Difference," in *Difference and Repetition*, 214–79.

40. Williams cited in Zukofsky, "Sincerity and Objectification," 198 (ellipses in the original text by Williams).

41. Zukofsky, "Sincerity and Objectification," 198.

42. Ibid., 199.

43. Williams, *Spring and All*, 183.

44. Tiffany, *Toy Medium*, 195.

45. Ibid., 171.

46. Bachelard, *The Philosophy of No*, 6.

47. Ibid., 6.

48. Ibid., 119.

49. See Bachelard, *Le rationalisme appliqué*. An excerpt from the third chapter of that book has been translated by Tiles as "Corrationalism and the Problematic," *Radical Philosophy* 173 (May/June 2012): 27–32.

50. Wang, "The Politics of Error."

51. Badiou, *The Century*, 97.

52. Oppen, "Route," 202.

CONCLUSION: *TECHNĒ*, *POIĒSIS*, FABRICATION

1. Ozin and Arsenault, *Nanochemistry*, xxx.

2. Stiegler, *Technics and Time, 1*, 137, 141.

3. Ibid., 142, 134.

4. Tiffany, *Toy Medium*, 291.

5. Drexler, "Machines of Inner Space," 326.

6. Whitehead, *Process and Reality*, 28.

7. Ibid., 214–15.

8. Althusser, "Contradiction and Overdetermination," 113.

9. Olson, *Maximus Poems*, III.90.

WORKS CITED

Agamben, Giorgio. *The Coming Community*. Translated by Michael Hardt. Minneapolis: University of Minnesota Press, 1993.

——. *Homo Sacer: Sovereign Power and Bare Life*. Translated by Daniel Heller-Roazen. Stanford, CA: Stanford University Press, 1998.

——. *The Open: Man and Animal*. Translated by Kevin Attell. Stanford, CA: Stanford University Press, 2004.

Ajayan, Pulickel M., and James M. Tour. "Nanotube Composites." *Nature* 447 (June 28, 2007): 1066–68.

Albright, Daniel. *Quantum Poetics: Yeats, Pound, Eliot, and the Science of Modernism*. Cambridge: Cambridge University Press, 1997.

Althusser, Louis. "Contradiction and Overdetermination." In *For Marx*. Translated by Ben Brewster. London: Verso, 2005.

Altieri, Charles. *Enlarging the Temple: New Directions in American Poetry During the 1960s*. Cranbury, NJ: Associated University Presses, 1979.

Applewhite, E .J. "The Naming of the Buckminsterfullerene." In *Buckminster Fuller: Anthology for the New Millennium*, edited by Thomas T. K. Zung. New York: St. Martin's Press, 2001.

AzoNano. "Nanotube Production Capacities and Demand." February 17, 2004. http://www.azonano.com/details.asp?ArticleID=511 (accessed April 28, 2016).

Bachelard, Gaston. "Corrationalism and the Problematic." Translated by Mary Tiles. *Radical Philosophy* 173 (May/June 2012): 27–32.

——. *Le matérialisme rationnel*. Paris: Presses Universitaires de France, 1953.

——. *The New Scientific Spirit*. Translated by Arthur Goldhammer. Boston: Beacon Press, 1984.

——. *The Philosophy of No*. Translated by G. C. Waterston. New York: Orion, 1968.

——. *Le rationalisme appliqué*. Paris: Presses Universitaires de France, 1949.

Badiou, Alain. *Being and Event*. Translated by Oliver Feltham. London: Continuum, 2005.

——. *The Century*. Translated by Alberto Toscano. Cambridge: Polity, 2007.

Baggott, Jim. *Perfect Symmetry: The Accidental Discovery of Buckminsterfullerene*. Oxford: Oxford University Press, 1994.

Bairoch, Paul. *Diagnostic de l'evolution economique du Tiers-Monde, 1900–1966*. Paris: Gauthier-Villars, 1967.

Ball, Philip. *Designing the Molecular World: Chemistry at the Frontier*. Princeton, NJ: Princeton University Press, 1994.

——. "Fullerenes Finally Score as Nobel Committee Honors Chemists." *Nature* 383 (October 17, 1996): 561.

——. "It's a Small World." *Chemistry World* 1.2 (February 2004). http://www.rsc.org/chemistryworld/restricted/2004/February/smallworld.asp (accessed April 28, 2016).

——. *Made to Measure: New Materials for the 21st Century*. Princeton, NJ: Princeton University Press, 1997.

——. "Roll Up for the Revolution." *Nature* 414 (November 8, 2001): 142–44.

——. "Round 'Em Up." *Nature* 365 (October 14, 1993): 606.

Banerjee, Sarbajit, Luisa Whittaker, and Christopher J. Patridge, "Microscopic and Nanoscale Perspective of the Metal Insulator Phase Transitions of VO2: Some New Twists to an Old Tale." *Journal of Physical Chemistry Letters* 2 (2011): 745–58.

Barthes, Roland. *The Rustle of Language*. Translated by Richard Howard. New York: Hill and Wang, 1986.

Baudrillard, Jean. *The Ecstasy of Communication*. Translated by Bernard Schutze and Caroline Schutze. Los Angeles: Semiotext(e), 1988.

Baughman, R. H., A. A. Zakhidov, and W. A. de Heer. "Carbon Nanotubes: The Route Toward Applications." *Science* 297 (August 2, 2002): 787–92.

Beach, Christopher. *ABC of Influence: Ezra Pound and the Remaking of American Poetic Tradition*. Berkeley: University of California Press, 1992.

Bellmer, Hans. "Memories of the Doll Theme." Translated by Peter Chametzky, Susan Felleman, and Jochen Schindler. *Sulfur* 26 (Spring 1990): 29–33.

Benanav, Aaron, and *Endnotes*. "Misery and Debt: On the Logic and History of Surplus Populations and Surplus Capital." *Endnotes* 2 (April 2010): 20–51.

Benfey, O. T. "August Kekulé and the Birth of the Structural Theory of Organic Chemistry in 1858." *Journal of Chemical Education* 35 (1958): 21–23.

Bennett, Jane. *Vibrant Matter: A Political Ecology of Things*. Durham, NC: Duke University Press, 2010.

Bensaude-Vincent, Bernadette, and Isabelle Stengers. *A History of Chemistry*. Translated by Deborah van Dam. Cambridge, MA: Harvard University Press, 1996.

Benveniste, Emile. *Problems in General Linguistics*. Translated by Mary Elizabeth Meek. Coral Gables: University of Miami Press, 1971.

Benyus, Janine M. *Biomimicry: Innovation Inspired by Nature*. New York: Harper Perennial, 1997.

Bergson, Henri. *Creative Evolution*. Translated by Arthur Mitchell. New York: Dover, 1998.

Bergvall, Caroline. *Fig.* Cambridge: Salt, 2005.

——. *Goan Atom.* San Francisco: Krupskaya, 2001.

——. "Speaking in Tongues: Caroline Bergvall in Conversation with John Stammers." By John Stammers. *Jacket* 22 (May 2003), http://jacketmagazine.com/22/bergv-stamm-iv.html (accessed April 28, 2016).

Blaser, Robin. "The Practice of Outside." In *The Fire: The Collected Essays of Robin Blaser*, edited by Miriam Nichols. Berkeley: University of California Press, 2006.

——. "The Violets: Charles Olson and Alfred North Whitehead." In *The Fire: The Collected Essays of Robin Blasé*, edited by Miriam Nichols. Berkeley: University of California Press, 2006.

Bohn, Willard. *The Aesthetics of Visual Poetry, 1914–1928.* Chicago: University of Chicago Press, 1993.

Bök, Christian. *Crystallography.* 2nd ed., rev. Toronto: Coach House, 2003.

——. *'Pataphysics: The Poetics of an Imaginary Science.* Evanston, IL: Northwestern University Press, 2002.

Boltanski, Luc, and Eve Chiapello. *The New Spirit of Capitalism.* Translated by Gregory Elliott. London: Verso, 2005.

Brassier, Ray. "Concept and Object." In *The Speculative Turn*, edited by Levi Bryant, Nick Srnicek, and Graham Harman. Melbourne: re.press, 2011.

Brown, Bill. "All Thumbs." *Critical Inquiry* 30.2 (Winter 2004): 452–57.

Brown, Nathan. "The Function of Digital Poetry at the Present Time." *Electronic Literature: New Horizons for the Literary.* http://newhorizons.eliterature.org/essay.php?id=11 (accessed April 28, 2016).

——. "Immortality ~~By Design~~." *Parallax* 48 (July–September 2008): 4–20.

——. "The Nadir of OOO: From Graham Harman's *Tool-Being* to Timothy Morton's *Realist Magic: Objects, Ontology, Causality.*" *Parrhesia* 17 (2013): 62–71.

——. "Needle on the Real: Technoscience and Poetry at the Limits of Fabrication." In *Nanoculture: Implications of the New Technoscience*, edited by N. Katherine Hayles. Bristol: Intellect Books, 2004.

——. "Next to Now." Review of Colin Milburn, *Nanovision: Engineering the Future. Radical Philosophy* 155 (May/June 2009): 57–59.

——. "Speculation at the Crossroads." Review of Tristan Garcia, *Form and Object: A Treatise on Things. Radical Philosophy* 188 (November/December 2014): 47–50.

Butler, Judith. *Bodies That Matter: On the Discursive Limits of "Sex."* New York: Routledge, 1993.

——. *Gender Trouble: Feminism and the Subversion of Identity.* New York: Routledge, 1999.

Byrd, Don. *Charles Olson's Maximus.* Urbana: University of Illinois Press, 1980.

——. "Cybernetics and Form." *OlsonNow.* http://epc.buffalo.edu/authors/olson/blog/ (accessed April 28, 2016).

———. *The Poetics of the Common Knowledge.* Albany: SUNY Press, 1994.
Cairns-Smith, A. G. *Genetic Takeover and the Mineral Origins of Life.* Cambridge: Cambridge University Press, 1982.
———. *Life's Puzzle: On Crystals and Organisms and on the Possibility of a Crystal as an Ancestor.* Edinburgh: Oliver and Boyd, 1971.
———. "The Origin of Life and the Nature of the Primitive Gene." *Journal of Theoretical Biology* 10 (January 1966): 53–88.
———. *Seven Clues to the Origin of Life: A Scientific Detective Story.* Cambridge: Cambridge University Press, 1985.
Cairns-Smith, A. G., and H. Harman, eds. *Clay Minerals and the Origin of Life.* Cambridge: Cambridge University Press, 1986.
Callister, William D., and David G. Rethwisch. *Materials Science and Engineering: An Introduction.* New York: John Wiley and Sons, 2010.
Cameron, Susan. *Choosing Not Choosing: Dickinson's Fascicles.* Chicago: Chicago University Press, 1992.
Campbell, Timothy. *Improper Life: Technology and Biopolitics from Heidegger to Agamben.* Minneapolis: University of Minnesota Press, 2011.
Cao Guozhong. *Nanostructures and Nanomaterials: Synthesis, Properties, and Applications.* London: Imperial College Press, 2004.
Chen, C. Julian. *Introduction to Scanning Tunneling Microscopy.* New York: Oxford University Press, 1993.
Clarke, Bruce, and Mark B. N. Hansen, eds. *Emergence and Embodiment: New Essays on Second-Order Systems Theory.* Durham, NC: Duke University Press, 2009.
Clover, Joshua. "Interview with Joshua Clover." By Gopal Balakrishnan. *Electronic Poetry Review* 8 (2008). http://www.epoetry.org/issues/issue8/text/prose/interview.htm (accessed April 28, 2016).
Cohen, Tom, et al. *Material Events: Paul de Man and the Afterlife of Theory.* Minneapolis: University of Minnesota Press, 2001.
Conte, Joseph. *Unending Design: The Forms of Postmodern Poetry.* Ithaca, NY: Cornell University Press, 1991.
Corman, Cid, ed. *Origin* 2 (Summer 1951).
Crommie, M. F., C. P. Lutz, and D. M. Eigler. "Confinement of Electrons to Quantum Corrals on a Metal Surface." *Science* 262 (October 8, 1993): 218–20.
———. "Imaging Standing Waves in a Two-Dimensional Electron Gas." *Nature* 363 (1993): 524–27.
Cuberes, M. T., R. R. Schittler, and J. K. Gimzewski. "Room-Temperature Repositioning of C60 Molecules at Cu Steps: Operation of a Molecular Counting Device." *Applied Physics Letters* 69.20 (1996): 3016–18.
Daston, Lorraine, and Peter Galison. *Objectivity.* New York: Zone Books, 2007.

Davenport, Guy. Introduction to *Valley of the Many-Colored Grasses* by Ronald Johnson, 9–14. New York: Norton, 1969.

Davidson, Michael. *Ghostlier Demarcations: Modern Poetry and the Material Word.* Berkeley: University of California Press, 1997.

——. *Guys Like Us: Citing Masculinity in Cold War Poetics.* Chicago: Chicago University Press, 2003.

Davies, Paul. Foreword to *Schrödinger's Machines: The Quantum Reshaping of Everyday Life*, by Gerard Milburn, vi–vii. New York: W. H. Freeman, 1997.

Dawkins, Richard. *The Blind Watchmaker.* London: Longmans, 1986.

Debord, Guy. *The Society of the Spectacle.* Translated by Donald Nicholson-Smith. New York: Zone Books, 1995.

Deleuze, Gilles. *Cinema 2: The Time Image.* Translated by Hugh Tomlinson and Robert Galeta. Minneapolis: University of Minnesota Press, 1989.

——. *Difference and Repetition.* Translated by Paul Patton. New York: Continuum, 2004.

——. *Différence et Répétition.* Paris: Presses Universitaires de France, 1968.

——. *Francis Bacon: The Logic of Sensation.* Translated by Daniel W. Smith. Minneapolis: University of Minnesota Press, 2003.

Dennett, Daniel. *Darwin's Dangerous Idea: Evolution and the Meanings of Life.* New York: Simon and Schuster, 1995.

Derrida, Jacques. "Signature Event Context." In *Margins of Philosophy.* Translated by Alan Bass. Chicago: University of Chicago Press, 1982.

Dewdney, Christopher. *A Paleozoic Geology of London, Ontario.* Toronto: Coach House, 1973.

——. "Parasite Maintenance." *Open Letter* 6&7 (Winter 1980–81): 19–37.

Dickinson, Emily. *The Manuscript Books of Emily Dickinson.* Edited by R.W. Franklin. Vol. 2. Cambridge, MA: Belknap Press, 1981.

——. *The Poems of Emily Dickinson: Reading Edition.* Edited by R.W. Franklin. Cambridge, MA: Belknap Press, 1999.

Dove, Patricia M., James J. De Yoreo, and Steve Weiner, eds. "Biomineralization." Special Issue, *Reviews in Mineralogy & Geochemistry* 54 (2003).

Doyle, Richard. *On Beyond Living: Rhetorical Transformations of the Life Sciences.* Stanford, CA: Stanford University Press, 1997.

——. *Wetwares: Experiments in Postvital Living.* Minneapolis: University of Minnesota Press, 2003.

Dresselhaus, M. S. "Down the Straight and Narrow." *Nature* 358 (July 16, 1992): 195–96.

Drexler, K. Eric. *The Engines of Creation: The Coming Era of Nanotechnology.* New York: Doubleday, 1986.

——. "Machines of Inner Space." In *Nanotechnology Research and Perspectives:*

Papers form the First Foresight Conference on Nanotechnology, edited by B. C. Crandall and James Lewis. Cambridge, MA: MIT Press, 1992.

Drucker, Johanna. *Experimental—Visual—Concrete: Avant-Garde Poetry Since the 1960s*. Amsterdam: Editions Rodopi, 1996.

———. *The Visible Word: Experimental Typography and Modern Art, 1909–1923*. Chicago: University of Chicago Press, 1997.

Duncan, Robert. "A Poem Beginning with a Line by Pindar." In *The Opening of the Field*. New York: New Directions, 1960.

———. "The Structure of Rime I." In *The Opening of the Field*. New York: New Directions, 1960.

DuPlessis, Rachel Blau. *Blue Studios: Poetry and Its Cultural Work*. Tuscaloosa: University of Alabama Press, 2006.

———. "Echological Scales: On *ARK* of Ronald Johnson." *Facture* 1 (2000): 99–119.

———. "Manhood and Its Poetic Projects: The Construction of Masculinity in the Counter-cultural Poetry of the U.S. 1950s." *Jacket* 31 (October 2006). http://jacket-magazine.com/31/duplessis-manhood.html (accessed April 28, 2016).

———. "Manifests." *Diacritics* 26.3/4 (Autumn–Winter 1996): 31–53.

Dworkin, Craig. *Reading the Illegible*. Evanston, IL: Northwestern University Press, 2003.

———. *Strand*. New York: Roof Books, 2005.

Ebbesen, T. W., and P. M. Ajayan. "Large-Scale Synthesis of Carbon Nanotubes." *Nature* 358 (July 16, 1992): 220–22.

Eigler, Donald. "From the Bottom Up: Building Things with Atoms." In *Nanotechnology*, edited by Gregory Timp. New York: Springer-Verlag, 1999.

Eigler, D. M., C. P. Lutz, and W. E. Rudge. "An Atomic Switch Realized with the Scanning Tunneling Microscope." *Nature* 352 (1991): 600–602.

Eigler, D. M., and E. K. Schweizer. "Positioning Single Atoms with a Scanning Tunneling Microscope." *Nature* 344 (1990): 524–26.

Emerson, Ralph Waldo. *The Essential Writings of Ralph Waldo Emerson*. Edited by Brooks Atkinson. New York: Modern Library, 2000

———. "Nature." In *The Essential Writings of Ralph Waldo Emerson*, edited by Brooks Atkinson. New York: Modern Library, 2000.

———. "The Poet." In *The Essential Writings of Ralph Waldo Emerson*, edited by Brooks Atkinson. New York: Modern Library, 2000.

———. "The Transcendentalist." In *The Essential Writings of Ralph Waldo Emerson*, edited by Brooks Atkinson. New York: Modern Library, 2000.

Esposito, Roberto. *Bíos: Biopolitics and Philosophy*. Translated by Timothy Campbell. Minneapolis: University of Minnesota Press, 2008.

Farr, Judith. *The Passion of Emily Dickinson*. Cambridge, MA: Harvard University Press, 1992.

Feynman, Richard. "There's Plenty of Room at the Bottom." (1959). In *Nanotechnology Research and Perspectives: Papers from the First Foresight Conference on Nanotechnology*, edited by B. C. Crandall and James Lewis. Cambridge, MA: MIT Press, 1992.

Finch, Annie. *The Ghost of Meter: Culture and Prosody in American Free Verse*. Ann Arbor: University of Michigan Press, 2000.

Folk, R. L. "Nanobacteria in Carbonate Sediments and Rocks." *Journal of Sedimentary Petrology* 63 (1993): 990–99.

Foster, J. S., J. E. Frommer, and P. C. Arnett. "Molecular Manipulation Using a Scanning Tunneling Microscope." *Nature* 331 (1988): 324–26.

Foucault, Michel. *Society Must Be Defended: Lectures at the College de France, 1975–1976*. Translated by David Macey. New York: Picador, 2003.

Fuller, R. Buckminster. "Conceptuality of Fundamental Structures." In *Structure in Art and Science*, edited by Gyorgy Kepes. New York: G. Braziller, 1965.

——. *Critical Path*. New York: St. Martin's Press, 1981.

——. *Intuition*. New York: Doubleday, 1972.

——. *Inventions: The Patented Work of R. Buckminster Fuller*. New York: St. Martin's, 1983.

——. *No More Secondhand God and Other Writings*. Carbondale: Southern Illinois University Press, 1963.

——. *Synergetics: Explorations in the Geometry of Thinking*. New York: Macmillan, 1975.

——. *Untitled Epic Poem on the History of Industrialization*. Highlands, NC: Jonathan Williams, 1962.

Garcia, Tristan. *Form and Object: A Treatise on Things*. Translated by Mark Allen Ohm and Jon Cogburn. Edinburgh: Edinburgh University Press, 2014.

Gene Ontology Annotation Database. http://www.ebi.ac.uk/GOA (accessed April 28, 2016).

Gille, Bertrand, ed. *Histoire des Techniques*. Paris: Gallimard, 1978.

Gimzewski, James K. "Nanotechnology: The Endgame of Materialism." *Leonardo* 41.3 (2008): 259–65.

Gimzewski, James, and Victoria Vesna. "The Nanomeme Syndrome: Blurring of Fact and Fictionin the Construction of a New Science." *Technoetic Arts: A Journal of Speculative Research* 1.1 (2003): 7–24.

Gimzewski, J. K., and M. E. Welland, eds. *Ultimate Limits of Fabrication and Measurement*. New York: Springer, 1995.

Glusker, Jenny P., and Kenneth N. Trueblood. *Crystal Structure Analysis: A Primer*. Oxford: Oxford University Press, 1985.

Goodsell, David S. *Bionanotechnology: Lessons from Nature*. Hoboken, NJ: Wiley Liss, 2004.

Graham, Jorie. "Pietà." In *The End of Beauty*. Hopewell, NJ: Ecco Press, 1987.
Grusin, Richard. *The Nonhuman Turn*. Minneapolis: University of Minnesota Press, 2015.
Hacking, Ian. *Representing and Intervening: Introductory Topics in the Philosophy of Natural Science*. Cambridge: Cambridge University Press, 1983.
——. *The Social Construction of What?* Cambridge, MA: Harvard University Press, 1999.
Haraway, Donna. *Modest_Witness@Second_Millennium.FemaleMan©_Meets_OncoMouse*TM. New York: Routledge, 1997.
Harman, Graham. "On Vicarious Causation." *Collapse* 2 (March 2007): 171–205.
——. *Tool-Being: Heidegger and the Metaphysics of Objects*. Chicago: Open Court, 2002.
Harris, Mary Emma. *The Arts at Black Mountain College*. Cambridge, MA: MIT Press, 1987.
Harris, Peter J. F. *Carbon Nanotubes and Related Structures: New Materials for the 21st Century*. Cambridge: Cambridge University Press, 2002.
Hartman, Geoffrey. *Criticism in the Wilderness: The Study of Literature Today*. New Haven, CT: Yale University Press, 1980.
Hawkins, Gary. "Constructing and Residing in the Paradox of Dickinson's Prismatic Space." *Emily Dickinson Journal* 9.1 (2000): 49–70.
Hayles, N. Katherine. *How We Became Posthuman: Virtual Bodies in Cybernetics, Literature, and Informatics*. Chicago: University of Chicago Press, 1999.
——, ed. *Nanotechnology: Implications of the New Technoscience*. Bristol: Intellect, 2004.
Hegel, G.F.W. *Phenomenology of Spirit*. Translated by A. V. Miller. Oxford: Oxford University Press, 1977.
Heidegger, Martin. *Being and Time*. Translated by John Macquarrie and Edward Robinson. San Francisco: HarperCollins, 1962.
——. *The Fundamental Concepts of Metaphysics: World, Finitude, Solitude*. Translated by William McNeill and Nicholas Walker. Bloomington: Indiana University Press, 1995.
——. ". . . Poetically Man Dwells . . ." In *Poetry, Language, Thought*. Translated by Albert Hofstadter. New York: HarperCollins, 1971.
——. "The Question Concerning Technology." In *The Question Concerning Technology and Other Essays*. Translated by William Lovin. New York: Harper and Row, 1977.
Heller, E. J., M. F. Crommie, C. P. Lutz, and D. M. Eigler. "Scattering and Absorption of Surface Electron Waves in Quantum Corrals." *Nature* 369 (July 7, 1994): 464–66.
Hill, Lindsay, and Paul Naylor, eds. *Facture* 1 (2000).

Hoeynck, Joshua. *The Principle of Measure in Composition by Field: Projective Verse II*. Victoria, TX: Chax, 2010.

Howe, Susan. *The Birth Mark: Unsettling the Wilderness in American Literary History*. Hanover, NH: Wesleyan University Press, 1993.

——. *My Emily Dickinson*. Berkeley: North Atlantic, 1985.

Hsu, Charlotte. "Working Toward 'Smart Windows.'" *UB Reporter* (14 April 2011) www.buffalo.edu/ubreporter/2011_04_14/smart_windows (accessed April 28, 2016).

Hui, Wang. "The 1989 Social Movement and the Historical Roots of China's Neoliberalism." Trans. Theodore Huters. In *China's New World Order: Society, Politics, and Economy in Transition*, edited by Theodore Huters. Cambridge, MA: Harvard University Press, 2003.

Hunt, Geoffrey, and Michael Mehta, eds. *Nanotechnology: Risk, Ethics, and Law*. London: Earthscan, 2006.

Iijima, Sumio. "Helical Microtubules of Graphitic Carbon." *Nature* 354 (November 7, 1991): 56–58.

Iles, Anthony, and Marina Vishmidt. "Work, Work Your Thoughts, and Therein See a Siege." In *Communization and Its Discontents: Contestation, Critique, and Contemporary Struggles*, edited by Benjamin Noyes. New York: Minor Compositions, 2011.

Jackson, Virginia. *Dickinson's Misery: A Theory of Lyric Reading*. Princeton, NJ: Princeton University Press, 2005.

Jakobsen, Roman. *Selected Writings*. Vol. 2. The Hague: Mouton, 1971.

Jameson, Fredric. *Postmodernism, or The Cultural Logic of Late Capitalism*. Durham, NC: Duke University Press, 1991.

Johnson, Ronald. *ARK*. Albuquerque: Living Batch, 1996.

——. *ARK 50: Spires 34–50*. New York: Dutton, 1984.

——. *The Book of the Green Man*. New York: Norton, 1967.

——. "From the Notebooks." Edited by Peter O'Leary. *Facture* 1 (2000): 59–97.

——. "Hurrah for Euphony." *Chicago Review* 42.1 (1996): 25–31.

——. "An Interview with Ronald Johnson." By Peter O'Leary. *Chicago Review* 42.1 (1996): 32–53.

——. "Planting the Rod of Aaron." *Northern Lights* 2 (1985–86): 1–14.

——. *RADI OS*. Berkeley: Sand Dollar, 1977.

——. *Valley of the Many-Colored Grasses*. New York: Norton, 1969.

Jones, Richard A.L. *Soft Machines: Nanotechnology and Life*. Oxford: Oxford University Press, 2004.

Juhasz, Suzanne. *The Undiscovered Continent: Emily Dickinson and the Space of the Mind*. Bloomington: Indiana University Press, 1983.

Jung, Yung Joon, et al. "Aligned Carbon Nanotube-Polymer Hybrid Architectures for Diverse Flexible Electronic Applications." *Nano Letters* 6.3 (2006): 413–18.

Katz, Vincent, ed. *Black Mountain College: Experiment in Art.* Cambridge, MA: MIT Press, 2002.

Kekulé, F. A. "Sur la constitution des substances aromatiques." *Bulletin de la Societe Chimique de Paris* 3 (1865): 98–110.

Kelleher, Michael, and Ammiel Alcalay. *OlsonNow.* http://olsonnow.blogspot.com/ (accessed April 11, 2007).

Kemp, Martin. "Kroto and Charisma." *Nature* 394 (July 30, 1998): 429.

Kenner, Hugh. *Bucky: A Guided Tour of Buckminster Fuller.* New York: William Morrow, 1973.

——. *Geodesic Math and How to Use It.* Berkeley: University of California Press, 2003.

——. *The Pound Era.* Berkeley: University of California Press, 1971.

Kher, Inder Nath. *The Landscape of Absence.* New Haven, CT: Yale University Press, 1974.

Kloeppel, James E. "DNA-wrapped Carbon Nanotubes Serve as Sensors in Living Cells" (January 26, 2006). https://news.illinois.edu/blog/view/6367/207077 (accessed April 28, 2016).

Kohn, Denise. "'I cannot live with You –'." In *An Emily Dickinson Encyclopedia,* edited by Jane Donahue Eberwein. Westport, CT: Greenwood, 1998.

Koyama, Yukinori, et al. "Harnessing the Actuation Potential of Solid-State Intercalation Compounds." *Advanced Functional Materials* 16 (2006): 492–98.

Krausse, Joachim, and Claude Lichtenstein, eds. *Your Own Private Sky: R. Buckminster Fuller.* Baden: Lars Müller, 1999.

Kristeva, Julia. *Revolution in Poetic Language.* Translated by Margaret Waller. New York: Columbia University Press, 1984.

Kroto, H. W. "Macro-, Micro-, and Nano-scale Engineering." In *Buckminster Fuller: Anthology for the New Millennium,* edited by Thomas T. K. Zung. New York: St. Martin's, 2001.

——. "The Stability of the Fullerenes C*n*, with n = 24, 28, 32, 36, 50, 60 and 70." *Nature* 329 (October 8, 1987): 529–31.

Kroto, H. W., J. R. Heath, S. C. O'Brien, R. F. Curl, and R. E. Smalley, "C60: Buckminsterfullerene." *Nature* 318 (November 14, 1985): 162–63.

Lacan, Jacques. *Seminar XX: On Feminine Sexuality, The Limits of Love and Knowledge, 1972–1973.* Translated by Bruce Fink. New York: Norton, 1988.

Latour, Bruno. *Reassembling the Social: An Introduction to Actor-Network-Theory.* Oxford: Oxford University Press, 2005.

——. *We Have Never Been Modern.* Translated by Catherine Porter. Cambridge, MA: Harvard University Press, 1993.

Lazzarato, Maurizio. "Immaterial Labor." Translated by Paul Colilli and Ed Emory. In *Radical Thought in Italy*, edited by Paolo Virno and Michael Hardt. Minneapolis: University of Minnesota Press, 1996.

Lee, Keekok. *Philosophy and Revolutions in Genetics: Deep Science and Deep Technology*. New York: Palgrave Macmillan, 2003.

Leroi-Gourhan, André. *Gesture and Speech*. Translated by Anna Bostok Berger. Cambridge, MA: MIT Press, 1993.

——. *L'homme et la matière*. Paris: Albin Michel, 1943.

Levertov, Denise. "Some Notes on Organic Form." In *New and Selected Essays*. New York: New Directions, 1992.

Lieber, Charles. "Scanning Tunneling Microscopy." *Chemical & Engineering News* 72 (April 18, 1994): 28–43.

Lin, Tan. *BlipSoak01*. Berkeley: Atelos, 2003.

Lippit, Akira. *Electric Animal: Toward a Rhetoric of Wildlife*. Minneapolis: University of Minnesota Press, 2000.

Luhmann, Niklas. *Social Systems*. Translated by John Bednarz and Dirk Baeker. Stanford, CT: Stanford University Press, 1996.

Mackey, Nathaniel. *Discrepant Engagement: Dissonance, Cross-Culturality, and Experimental Writing*. Tuscaloosa: University of Alabama Press, 1993.

Maddox, Brenda. *Rosalind Franklin: The Dark Lady of DNA*. New York: Harper Perennial, 2003.

Magner, Lois N. *A History of the Life Sciences*. New York: Marcel Dekker, 1979.

Malabou, Catherine. *Plasticity at the Dusk of Writing: Dialectic, Destruction, Deconstruction*. Translated by Caroline Shread. New York: Columbia University Press, 2010.

——. *What Should We Do with Our Brain?* Translated by Sebastian Rand. New York: Fordham, 2008.

Mandel, Ernest. *Late Capitalism*. Translated by Joris De Bres. London: NLB, 1975.

Maniloff, J. "Nannobacteria: Size Limits and Evidence." *Science* 276 (1997): 1773–76.

Marks, Robert, and Buckminster Fuller. *The Dymaxion World of Buckminster Fuller*. New York: Anchor Books, 1973.

Martin, Reinhold. *The Organizational Complex: Architecture, Media, and Corporate Space*. Cambridge, MA: MIT Press, 2003.

Marx, Karl. *Capital Volume 1*. Translated by Ben Fowkes. New York: Penguin, 1976.

Maturana, Humberto, Evan Thompson, and Eleanor Rusch. *The Embodied Mind: Cognitive Science and Human Experience*. Cambridge, MA: MIT Press, 1993.

Maturana, Humberto and Francisco Varela. *Autopoiesis and Cognition: The Realization of the Living*. Dordrecht: Reidel, 1980.

——. *The Tree of Knowledge: The Biological Roots of Human Understanding*. Boston: Shambhala, 1987.

Mau, Bruce, et al. *Massive Change*. London: Phaidon, 2004.

McCaffery, Steve. "Carnival" (1975). In *Seven Pages Missing, Volume I*. Toronto: Coach House, 2000.

——. *Prior to Meaning: The Protosemantic and Poetics*. Evanston, IL: Northwestern University Press, 2001.

——. *Shifters* (1976). In *Seven Pages Missing, Volume One: Selected Texts, 1969–1999*. Toronto: Coach House, 2000.

——. "Shifters: A Note" (1976). In *Seven Pages Missing, Volume One: Selected Texts, 1969–1999*. Toronto: Coach House, 2000.

——. "Strata and Strategy: 'Pataphysics in the Poetry of Christopher Dewdney." In *North of Intention: Critical Writings, 1973–1986*. New York: Roof, 2000.

McCaffery, Steve, and bpNichol. *Rational Geomancy: The Kids of the Book Machine, The Collected Research Reports of the Toronto Research Group, 1973–1982*. Vancouver: Talon Books, 1992.

Michaels, Walter Benn. *The Shape of the Signifier*. Princeton, NJ: Princeton University Press, 2004.

Milburn, Colin. "Nano/Splatter: Disintegrating the Postbiological Body." *New Literary History* 36 (2005): 283–311.

——. "Nanotechnology in the Age of Posthuman Engineering: Science Fiction as Science." In *Nanoculture: Implications of the New Technoscience*, edited by N. Katherine Hayles. Bristol: Intellect Books, 2004.

——. *Nanovision: Engineering the Future*. Durham, NC: Duke University Press, 2008.

Milburn, Gerard. *Schrödinger's Machines: The Quantum Reshaping of Everyday Life*. New York: W. H. Freeman, 1997.

MillAr, Jay. *Mycological Studies*. Toronto: Coach House, 2002.

Miller, Christanne. *Emily Dickinson: A Poet's Grammar*. Cambridge, MA: Harvard University Press, 1987.

Milne, Drew. "A Veritable Dollmine: Caroline Bergvall, *Goan Atom*, 1: *jets poupee*." *Jacket* 12 (July 2000). http://jacketmagazine.com/12/milne-bergvall.html (accessed April 28, 2016).

Mossen, Andrew. "'In Thicket': Charles Olson, Frances Boldereff, Robert Creeley, and the Crisis of Masculinity at Mid-Century." *Modern Literature* 28.4 (2005): 13–39.

Nancy, Jean-Luc. *Corpus*. Translated by Richard A. Rand. New York: Fordham University Press, 2008.

——. "Of Being Singular Plural." In *Being Singular Plural*. Translated by Robert D. Richardson and Anne E. O'Byrne. Stanford, CT: Stanford University Press, 2000.

——. "Res ipsa et ultima." Translated by Steven Miller. In *A Finite Thinking*, edited by Simon Sparks. Stanford, CT: Stanford University Press, 2003.

——. *The Sense of the World*. Translated by Jeffrey S. Librett. Minneapolis: University of Minnesota Press, 1997.

——. "The Sublime Offering." Translated by Jeffrey S. Librett. In *A Finite Thinking*, edited by Simon Sparks. Stanford, CT: Stanford University Press, 2003.

Nelson, Paul. "Organic Poetry." *OlsonNow*. http://epc.buffalo.edu/authors/olson/blog/ (accessed April 28, 2016).

Olson, Charles. "Against Wisdom as Such." In *Collected Prose: Charles Olson*, edited by Donald Allen and Benjamin Friedlander. Berkeley: University of California Press, 1997.

——. *Causal Mythology*. San Francisco: Four Seasons, 1969.

——.*Charles Olson & Robert Creeley: The Complete Correspondence*. Vol. 5. Edited by George F. Butterick. Santa Barbara, CA: Black Sparrow, 1983.

——. *Charles Olson & Robert Creeley: The Complete Correspondence*. Vol. 10. Edited by Richard Blevins. Santa Barbara, CA: Black Sparrow, 1996.

——. *Charles Olson in Connecticut: Last Lectures*, edited by John Cech, Oliver Ford, and Peter Rittner. Iowa City: Windhover Press, University of Iowa, 1974.

——. "The Chiasma, or Lectures in the New Sciences of Man." *Olson: The Journal of the Charles Olson Archives* 10: 3–109.

——. "In Cold Hell, In Thicket." In *The Distances*. New York: Grove, 1960.

——. "The Gate and the Center." In *Collected Prose: Charles Olson*, edited by Donald Allen and Benjamin Friedlander. Berkeley: University of California Press, 1997.

——. "Equal, That Is, to the Real Itself." In *Collected Prose: Charles Olson*, edited by Donald Allen and Benjamin Friedlander. Berkeley: University of California Press, 1997.

——. "Human Universe." In *Collected Prose: Charles Olson*, edited by Donald Allen and Benjamin Friedlander. Berkeley: University of California Press, 1997.

——. "The Kingfishers." In *The Distances*. New York: Grove, 1960.

——. *Letters for Origin: 1950–1956*. Edited by Albert Glover. New York: Paragon House, 1969.

——. "The Present is Prologue." In *Collected Prose: Charles Olson*, edited by Donald Allen and Benjamin Friedlander. Berkeley: University of California Press, 1997.

——. "Projective Verse." In *Collected Prose: Charles Olson*, edited by Donald Allen and Benjamin Friedlander. Berkeley: University of California Press, 1997.

——. *Proprioception*. In *Collected Prose: Charles Olson*, edited by Donald Allen and Benjamin Friedlander. Berkeley: University of California Press, 1997.

——. *The Maximus Poems*. Edited by George Butterick. Berkeley: University of California Press, 1983.

——. *Mayan Letters*. Edited by Robert Creeley. London: Jonathan Cape, 1953.

——. *The Special View of History*. Edited by Ann Charters. Berkeley: Oyez, 1970.

Olson, Gregory B. "Designing a New Material World." *Science* 288.5468 (May 2000): 993–98.

Oppen, George. *Of Being Numerous*. In *New Collected Poems*, edited by Michael Davidson. New York: New Directions, 2002.

———. "Route." In *New Collected Poems*, edited by Michael Davidson. New York: New Directions, 2002.

———. *The Selected Letters of George Oppen*. Edited by Rachel Blau DuPlessis. Durham, NC: Duke University Press, 1990.

Ostman, Charles. "The Nanobiology Imperative." www.historianofthefuture.com/nanobio.html (accessed April 28, 2016).

———. "Nanotechnology: The Next Revolution." *Inside: The Magazine of the 21st Century* (1997). http://web.archive.org/web/20000607223454/21net.com/content/inside_se/toc.htm (accessed April 28, 2016).

Ozin, Geoffrey A., and André C. Arsenault. *Nanochemistry: A Chemical Approach to Nanomaterials*. Cambridge: Royal Society of Chemistry, 2005.

Paul, Sherman. *Olson's Push: Origin, Black Mountain, and Recent American Poetry*. Baton Rouge: Louisiana State University Press, 1978.

Perelman, Bob. *Ten to One: Selected Poems*. Hanover, NH: Wesleyan University Press, 1999.

Perloff, Marjorie. "Charles Olson and the 'Inferior Predecessors': 'Projective Verse' Revisited." *ELH* 40.2 (Summer 1973): 285–306.

———. *The Dance of Intellect: Studies in the Pound Tradition*. Chicago: Northwestern University Press, 1996.

———. "The Oulipo Factor: The Procedural Poetics of Christian Bök and Caroline Bergvall." *Jacket* 23 (August 2003). http://jacketmagazine.com/23/perlof-oulip.html (accessed April 28, 2016).

———. *Radical Artifice: Writing in the Age of Media*. Chicago: Chicago University Press, 1991.

———. "Songs of the Earth: Ronald Johnson's Verbivocovisuals." In *Differentials: Poetry, Poetics, Pedagogy*. Tuscaloosa: University of Alabama Press, 2004.

Pound, Ezra. *The Cantos of Ezra Pound*. New York: New Directions, 1998.

———. *Personae*. Ed. Lea Baechler and A. Walton Litz. New York: New Directions, 1990.

———. "A Retrospect." In *Literary Essays of Ezra Pound*, edited by T. S. Eliot. New York: New Directions, 1968.

Prigogine, Ilya, and Isabelle Stengers. *Order Out of Chaos: Man's New Dialogue with Nature*. New York: Bantam, 1984.

Pynchon, Thomas. *Gravity's Rainbow*. New York: Penguin, 1973.

Rancière, Jacques. *The Politics of Aesthetics: The Distribution of the Sensible*. Translated by Gabriel Rockhill. London: Continuum, 2004.

Rasmussen, Steen, et al. "Transitions from Nonliving to Living Matter." *Science* 303.5660 (2004): 963–965.

Rasula, Jed. *This Compost: Ecological Imperatives in American Poetry.* Athens: University of Georgia Press, 2002.

——. "The Poetics of Embodiment: A Theory of Exceptions." PhD diss., University of California at Santa Cruz, 1989.

Rasula, Jed, and Steve McCaffery, eds. *Imagining Language: An Anthology.* Cambridge, MA: MIT Press, 2001.

Ratner, Mark, and Daniel Ratner. *Nanotechnology: A Gentle Introduction to the Next Big Idea.* Upper Saddle River, NJ: Prentice Hall, 2003.

RCSB Protein Data Bank. http://www.rcsb.org/pdb/statistics/holdings.do (accessed April 26, 2016).

Reed, Mark. "Quantum Constructions." *Science* 262 (1993): 195.

Regis, Ed. *Nano!: Remaking the World Atom by Atom.* London: Bantam, 1995.

Rilke, Ranier Maria. "The Eighth Duino Elegy." In *Ahead of All Parting: The Selected Poetry and Prose of Rainer Maria Rilke.* Translated by Stephen Mitchell. New York: Modern Library, 1995.

Rodes, Gale. *Crystallography Made Crystal Clear: A Guide for Users of Macromolecular Models.* 3rd ed. Burlington, MA: Academic Press, 2006.

Santner, Eric L. *On Creaturely Life: Rilke, Benjamin, Sebald.* Chicago: University of Chicago Press, 2006.

Schulman, Rebecca, and Erik Winfree. "How Crystals That Sense and Respond to Their Environments Could Evolve." *Natural Computing* 7 (2008): 219–37.

——. "Self-Replication and Evolution of DNA Crystals." *Lecture Notes in Computer Science* 3630 (2005): 734–43.

Scroggins, Mark. "'A' to *ARK*: Zukofsky, Johnson, and an Alphabet of the Long Poem." *Facture* 1 (2000): 143–52.

Sedgwick, Eve Kosofsky. *The Epistemology of the Closet.* Berkeley: University of California Press, 2008.

Seeman, Nadrian C. "Nanotechnology and the Double Helix." *Scientific American* 290.6 (2004): 65–75.

Sieden, Lloyd Steven. *Buckminster Fuller's Universe: An Appreciation.* New York: Plenum, 1989.

Simondon, Gilbert. *Du mode d'existence des objets techniques.* Paris: Aubier, 1958.

Smalley, Richard. "Future Global Energy Prosperity: The Terrawatt Challenge." *Materials Research Society Bulletin* 30 (June 2005): 412–17.

——. "Nanotechnology and the Next 50 Years." December 17, 1995. http://cohesion.rice.edu/naturalsciences/smalley/smalley.cfm?doc_id=5336 (accessed April 28, 2016).

Snyder, Gary. *Rivers and Mountains Without End.* San Francisco: North Point, 1996.

Sobchack, Vivian. *Carnal Thoughts: Embodiment and Moving Image Culture.* Berkeley: University of California Press, 2004.

Sohn-Rethel, Alfred. *Intellectual and Manual Labor: A Critique of Epistemology.* Trans. Martin Sohn-Rethel. London: Macmillan, 1978.

Solt, Mary Ellen, ed. *Concrete Poetry: A World View.* Bloomington: University of Indiana Press, 1970.

Stefans, Brian Kim. "Make It New." *Boston Review* (September/October 2006). https://www.bostonreview.net/brian-kim-stefans-make-it-new (accessed April 28, 2016).

Stengers, Isabelle. *Thinking with Whitehead: A Free and Wild Creation of Concepts.* Translated by Michael Chase. Cambridge, MA: Harvard University Press, 2011.

Stiegler, Bernard. *Technics and Time, 1: The Fault of Epimetheus.* Translated by Richard Beardsworth and George Collins. Stanford, CA: Stanford University Press, 1998.

Stonum, Gary Lee. *The Dickinson Sublime.* Madison: University of Wisconsin Press, 1990.

Strano, Michael S. et al., "Optical Detection of DNA Conformational Polymorphism on Single-Walled Carbon Nanotubes." *Science* 311.5760 (January 27, 2006): 508–11.

Taya, Minoru. "Bio-inspired Design of Intelligent Materials." *Smart Structures and Materials 2003: Electroactive Polymer Actuators and Devices* (July 2003): 54–65.

Taylor, Sue. *Hans Bellmer: The Anatomy of Anxiety.* Cambridge, MA: MIT Press, 2000.

Thacker, Eugene. *Biomedia.* Minneapolis: University of Minnesota Press, 2004.

———. *The Global Genome.* Cambridge, MA: MIT Press, 2005.

Thompson, D'Arcy. *On Growth and Form.* Edited by John Tyler Bonner. Cambridge: Cambridge University Press, 1961.

Tiffany, Daniel. *Radio Corpse: Imagism and the Cryptaesthetic of Ezra Pound.* Cambridge, MA: Harvard University Press, 1995.

———. *Toy Medium: Materialism and Modern Lyric.* Berkeley: University of California Press, 2000.

Timp, Gregory, ed. *Nanotechnology.* New York: Springer-Verlag, 1999.

Tomlin, Sarah. "Getting to Grips with Smart Materials." *Nature: Materials Update* (April 2002). http://www.nature.com/materials/news/news/020404/portal/m020404–4.html (accessed May 15, 2006).

Turberfield, Andrew. "DNA as Engineering Material." *Physics World* 16.3 (March 2003): 43–46.

Uwins, J. R., et al. "Novel Nano-Organism from Australian Sandstones." *American Mineralogist* 83 (1998): 1541–50.

Varela, Francisco, Evan Thompson, and Eleanor Rusch. *The Embodied Mind: Cognitive Science and Human Experience.* Cambridge, MA: MIT Press, 1991.

Vendler, Helen. *Part of Nature, Part of Us: Modern American Poets.* Cambridge, MA: Harvard University Press, 1980.

Virno, Paolo. *A Grammar of the Multitude: For an Analysis of Contemporary Forms of Life*. Translated by Isabelle Bertoletti, James Cascaito, and Andrea Casson. Los Angeles: Semiotexte, 2004.

von Hallberg, Robert. *Charles Olson: The Scholar's Art*. Cambridge, MA: Harvard University Press, 1978.

von Uexküll, Jakob. *A Foray into the Worlds of Animals and Humans*. Translated by Joseph D. O'Neil. Minneapolis: University of Minnesota Press, 2010.

Wang, Shanxing. *Mad Science in Imperial City*. New York: Futurepoem, 2005.

——. "The Politics of Error: An Interview with Shanxing Wang." By Nathan Brown. *Jacket* 34 (October 2007). http://jacketmagazine.com/34/brown-iv-shanxing.shtml (accessed April 28, 2016).

Wang, Zhong Lin, Yi Lui, and Ze Zhang. *Handbook of Nanophase and Nanostructured Materials*. Vol. 2. New York: Kluwer, 2003.

Watten, Barrett. *The Constructivist Moment: From Material Text to Cultural Poetics*. Middleton, CT: Wesleyan University Press, 2003.

——. "Radical Particularity, Critical Regionalism, and the Resistance to Globalization." (2007). http://www.english.wayne.edu/fac_pages/ewatten/posts/post42.html (accessed August 16, 2012).

Whitehead, Alfred North. *The Concept of Nature*. Amherst, NY: Prometheus Books, 2004.

——. *Process and Reality*. Edited by David Ray Griffin and Donald W. Sherburne. New York: Free Press, 1978.

——. *Science and the Modern World*. New York: Free Press, 1925.

Whitesides, George M., and Bartosz Grzybowski. "Self-Assembly at All Scales." *Science* 295 (March 29, 2002): 2418–21.

"Whither Crystallography." *Nature: Structural Biology* 8.11 (November 2001): 909.

Wiener, Norbert. *The Human Use of Human Beings: Cybernetics and Society*. New York: Da Capo, 1954.

Williams, William Carlos. *Spring and All*. In *The Collected Poems of William Carlos Williams: Volume I, 1909–1939*, edited by A. Walton Litz and Christopher MacGowan. New York: New Directions, 1986.

Wilson, Michael, et al. *Nanotechnology: Basic Science and Emerging Technologies*. New York: Chapman and Hall, 2002.

Wolff, Cynthia Griffen. *Emily Dickinson*. Reading, MA: Addison Wesley, 1988.

Wolfram, Stephen. *A New Kind of Science*. New York: Wolfram Media, 2002.

Yakobseon, Boris I., and Luise S. Couchman. "Carbon Nanotubes: Supramolecular Mechanics." In *Dekker Encyclopedia of Nanoscience and Nanotechnology*. Vol. 1. Edited by James A. Schwartz, Cristian I. Contescu, and Carol Putyera. New York: Dekker, 2004.

Yurke, Bernard, et al. "A DNA-Fuelled Molecular Machine Made of DNA." *Nature* 406 (August 2000): 605–8.

Žižek, Slavoj. *The Sublime Object of Ideology*. London: Verso, 1989.
Zukofsky, Louis. "A." Baltimore: Johns Hopkins University Press, 1978.
——. "An Objective." In *Prepositions +: The Collected Critical Essays*, edited by Mark Scroggins. Hanover, NH: Wesleyan University Press, 2000.
——. "Program: 'Objectivists' 1931." In *Prepositions +: The Collected Critical Essays*, edited by Mark Scroggins. Hanover, NH: Wesleyan University Press, 2000.
——. "Sincerity and Objectification." In *Prepositions +: The Collected Critical Essays*, edited by Mark Scroggins. Hanover, NH: Wesleyan University Press, 2000.

INDEX

IDIOM INVENTING WRITING THEORY

Jacques Lezra and Paul North, series editors

Werner Hamacher, *Minima Philologica*. Translated by Catharine Diehl and Jason Groves

Michal Ben-Naftali, *Chronicle of Separation: On Deconstruction's Disillusioned Love*. Translated by Mirjam Hadar

Daniel Hoffman-Schwartz, Barbara Natalie Nagel, and Lauren Shizuko Stone, eds., *Flirtations: Rhetoric and Aesthetics This Side of Seduction*

Márton Dornbach, *Receptive Spirit: German Idealism and the Dynamics of Cultural Transmission*

Jean-Luc Nancy, *Intoxication*. Translated by Philip Armstrong

Sean Alexander Gurd, *Dissonance: Auditory Aesthetics in Ancient Greece*

Anthony Curtis Adler, *Celebricities: Media Culture and the Phenomenology of Gadget Commodity Life*

Nathan Brown, *The Limits of Fabrication: Materials Science, Materialist Poetics*